Robert Lowe

General View of the Agriculture of the County of Nottingham

With Observations on the Means of Its Improvement

Robert Lowe

General View of the Agriculture of the County of Nottingham
With Observations on the Means of Its Improvement

ISBN/EAN: 9783741180682

Manufactured in Europe, USA, Canada, Australia, Japa

Cover: Foto ©Andreas Hilbeck / pixelio.de

Manufactured and distributed by brebook publishing software (www.brebook.com)

Robert Lowe

General View of the Agriculture of the County of Nottingham

GENERAL VIEW

OF THE

AGRICULTURE

OF THE

COUNTY

OF

NOTTINGHAM;

WITH

OBSERVATIONS ON THE MEANS OF ITS IMPROVEMENT.

DRAWN UP FOR THE CONSIDERATION OF

THE BOARD OF AGRICULTURE,

AND INTERNAL IMPROVEMENT.

BY ROBERT LOWE, ESQ.

OF OXTON.

LONDON:

PRINTED FOR RICHARD PHILLIPS, BRIDGE-STREET, BLACKFRIARS;

SOLD BY FAULDER & SON, BOND STREET; REYNOLDS, OXFORD STREET; J. HARDING, ST. JAMES'S STREET; J. ASPERNE, CORNHILL; BLACK, PARRY, & KINGSBURY, LEADENHALL STREET; BURBAGE & STRETTON, & SUTTON, & W. SMITH, NOTTINGHAM; TAYLOR, RETFORD; SHEPPARD, MANSFIELD; HAGE, NEWARK; CONSTABLE & CO. EDINBURGH; J. ARCHER, DUBLIN; & ALL OTHER BOOKSELLERS; BY D. M'MILLAN, BOW STREET, COVENT GARDEN.

1798.

[Price Five Shillings in Boards.]

CONTENTS.

CHAP. I.—*Geographical State and Circumstances.*

		PAGE
SECT. 1.	Situation and Extent.	
SECT. 2.	Divisions	
SECT. 3.	Climate	
	Dryness	2
SECT. 4.	Soil and Surface	
	Division of Districts according to soil,	3, 4, 5
SECT. 5.	Minerals and Fossils	5
	Stone	ib.
	Coals	ib.
	Lime	6
	Gypsum	ib.
	Marl	ib.
SECT. 6.	Water	ib.
	Rivers and Brooks	6, 7

CHAP. II.—*State of Property.*

SECT. 1.	Estates and their Management	8
SECT. 2.	Farms	ib.

CHAP. III.—*Buildings.*

Houses, Farm Houses, Cottages, &c.	9
Staddles	10
Fother Room	ib.
Particular method of laying barn floors	ib.

CONTENTS.

CHAP. IV.—*Mode of Occupation.*

		PAGE
SECT. 1.	Size of Farms, Character of Farmers	14
SECT. 2.	Rent in Money, Personal Services	ib.
SECT. 3.	Tythes	15
SECT. 4.	Poor's Rates	ib.
SECT. 5.	Leases	16
SECT. 6.	Expence and Profit	ib.

CHAP. V.—*Implements of Husbandry.*

Various Ploughs	17
Harrow	ib.
Threshing Machines	18
Waggons and Carts	ib.

CHAP. VI.—*Inclosing, Fences, Gates.*

Whitethorn Fence	19
Birch ditto	ib.
Gates Oak	ib.
——— Willow	ib.
Hedgerow Planting	ib.

CHAP. VII.—*Employment of Land, Arable Land, Cultivation, and Rotation of Crops.*

SECT. 1.	Employment of Land in Forest District	21
	Forest Breaks	ib.
	Clumber Park	22
	Newstead Park	23
	Beskwood Park	ib.
	Cultivation and Employment in Forest Inclosures	24
	Particular Articles in Forest District	27
	Weld	ib.
	Hops	ib
	Liquorice	28

CONTENTS.

	PAGE
Cultivation in Trent Bank District	28
Arable, course of	ib.
Winter Tares	29
Grass Lands	ib.
Tongue of Land east of Trent	ib.
Letter from G. Neville, Esq. of Thorney, on Cultivation, Stock, and Planting of the above Land	30
Paring and Burning	31
Lime	ib.
Whale Blubber	32
Sticklebacks	ib.
Sheep	ib.
Skegs	33
Planting	34
Uses of Birch	ib.
Occupation and Cultivation in Clay District North of Trent	37
Winter Tares	38
Hops	ib.
Pasture	42
Occupation in Vale of Belvoir District	44
Ditto in Nottinghamshire Woulds	ib.
Ditto in Lime and Coal District	45
Course of Crops on Limestone	ib.
Ditto on Coal Land	ib.

SECT. 2.

Crops commonly cultivated	ib.
Wheat	ib.
Rye	46
Barley	ib.
Oats	ib.
Skegs	ib.
Beans	ib.
Pease	ib.
Buckwheat	47

CONTENTS.

CHAP. VIII.—*Grass, Natural and Artificial.*

		PAGE
SECT. 1.	Red and Broad Clover	48
	Trefoil	ib.
	Rib Grass	ib.
	Burnet	ib.
	Sainfoin	ib.
	Lucerne	49
SECT. 2.	Hay-harvest	ib.

CHAP. IX.—*Gardens, Orchards, and Nurseries.*

CHAP. X.—*Woods and Plantations.*

In Forest District	51
Hays of Birkland and Bilhagh	ib.
Plantations	53
Method of Planting at Clumber	54
Account of Plantations on the Duke of Portland's Estate	57
Woods and Plantations in Forest District and Borders	70
Duke of Newcastle's Ash Plantations, and Value of Land	71
Ditto Plantations of Oak	73
Woods and Plantations of Hon. R. Lumley Saville	75
Ditto of Charles Pierrepont, Esq. now Lord Newark	78
Plantations of F. F. Foljambe, Esq.	80
Ditto of Charles Mellish, Esq.	ib.
Ditto of Taylor and Wollaston White, Esqrs.	ib.
Ditto of Joseph Cowlishaw	ib.
Woods and Plantations of D. Norfolk, at Worshop Manor	82

	PAGE
Woods in Trent Bank District	83
——— Lord Middleton's	84
——— Mr. Duncombe's	ib.
——— Roger Pocklington, Esq.	ib.
——— J. Pocklington, Esq.	ib.
Plantations at Thorney (vide Mr. Neville's Letter, p. 30)	
Woods in Clay District north of Trent	84
Thorney Wood Chace	85
General List of Woods and Plantations in Clay District north of Trent	86
Account of Ossingham Woods	89
Epperstone Woods	90
Woods in Thorney Wood Chace allotted, &c.	ib.
Vale of Belvoir and Woulds	91
Lime and Coal District	ib.
Lord Middleton's Woods	ib.
Further List of Woods in Lime and Coal District	92
Particulars of Mr. Roulston's Woods	ib.
——— of Lord Bathurst's	93
——— of Mr. Knight's	94
Soil of Woods	95

CHAP. XI.—*Wastes.*

Sherwood Forest	96

CHAP. XII.—*Improvements.*

Sect. 1.	Drainage	98
	Mr. Dixon's Letter on ditto	ib.
	Covered Drains	100
Sect. 2.	Watering	101
Sect. 3.	Manuring, Paring, and Burning	103
	Farm Yard Dung in Forest District	ib.

CONTENTS.

		PAGE
	Lime	104
	Dove Manure	108
	Bone Dust	109
	Rape Dust	ib.
	Green Manures	110
	Malt Combs	ib.
	Scrapings of Oiled Leather	ib.
	Bog Earth	ib.
	Gypsum or Plaister	ib.
	Skerry Stone Pounded	111
	Whale Blubber	ib.
	Soot	ib.
	Compost Dunghill	ib.
	Paring and Burning	112
Sect. 4.	Weeding	116
	Sheeping Beans	ib.
	Letter from Mr. Calvert, on Weeding	117
Sect. 5.	Planting	ib.
	Cultivation of Willows	118
	Fencing Inclosures	121
	Hedge-row Planting	122

CHAP. XIII.—*Live Stock.*

Sect. 1.	Black Cattle in the different Districts	123
Sect. 2.	Sheep in the different Districts	124
	Different Crosses in Breeding	125, 126
	Experiment in feeding different Breeds	125
Sect. 3.	Horses and their Use in Husbandry compared to Oxen	130
	Oxen used	ib.
Sect. 4.	Hogs	131
Sect. 5.	Rabbits	ib.
Sect. 6.	Poultry	ib.
Sect. 7.	Pigeons	132
	Remarkable Fact	ib.
Sect. 8.	Bees	ib.

CHAP. XIV.—*Rural Oeconomy.*

		PAGE
SECT. 1.	Labour, Servants, Labourers, Hours of Labour	133
SECT. 2.	Provisions	ib.
SECT. 3.	Fuel	134

CHAP. XV.—*Political Oeconomy as connected with Agriculture.*

SECT. 1.	Roads	135
SECT. 2.	River Navigation and Canals	136
SECT. 3.	Fairs and Markets	137
SECT. 4.	Commerce and Manufactures	138
SECT. 5.	Poor	140
SECT. 6.	Population	ib.

CHAP. XVI.—*Obstacles to Improvement.*

Execution of the Laws of Sewers defective	141
Queries as to Tythes	ib.

CHAP. XVII.—*Miscellaneous Observations.*

Agricultural Society	143
Weights and Measures	ib.

CONCLUSION, - - - - 144

APPENDIX.

No. 1.	Measure of Rain fallen at East Bridgford, &c.	145
No. 2.	Account of Skegs	
No. 3.	Statement of Cultivation and Improvement of Clumber Park	149
No. 4.	Private Inclosures from Forest and Borders	150

		PAGE
No. 5.	Account of the Cultivator	151
No. 6.	List of Open and Inclosed Townships, in the Clay District North of Trent	154
No. 7.	Ditto in the Vale of Belvoir and Notts Woulds	155
No. 8.	Extent, Jurisdiction, and Officers of Sherwood Forest	156
No. 9.	Letter from Mr. Calvert, on Miscellaneous Articles of Culture	158
No. 10.	Observations of Mr. Green, of Bankwood, on Reading the Survey of Mid Lothian and applying them to Notts	167
No. 11.	Cotton and Wool Mills in Nottingham and the County	170
No. 12.	Lists of Births, Burials and Inhabitants in the County of Nottinghamshire	172
	Paper B—Account of the number of Inhabitants in the Town of Nottingham, in 1779	179
No. 13.	Way of Making Ponds in dry Pastures	187
No. 14.	Cure of Diseases in Corn and Cattle	188
	Smut in Wheat	ib.
	Rot in Sheep	189
	The Water	ib.
No. 15.	Letter from Mr. Raynes, with Account of Cultivation of a Sand Farm	190

PRELIMINARY OBSERVATION.

THE Surveyor begs leave to premise, that in undertaking the work he proposed to himself only to state, as far as came to his knowledge, the usual course of husbandry used in the county, the new practices introduced, and such improvements as suggested themselves to him; without pretending to enter deep into scientific disquisition, or the subject of political regulation, which the reader must therefore expect to be but slightly touched upon.

AGRICULTURAL SURVEY OF NOTTINGHAMSHIRE.

CHAPTER I.
Geographical State and Circumstances.

SECTION I.
SITUATION AND EXTENT.

THE county of NOTTINGHAM is situated between fifty-two deg. fifty min. and fifty-three deg. thirty-four min. north latitude. It is about fifty miles in length, and twenty-five in breadth; and is supposed to contain about 480,000 acres. It has Derbyshire on the west, Yorkshire on the north, Lincolnshire on the east, and Leicestershire on the south.

SECT. II.—DIVISIONS.

This county is divided into six wapentakes or hundreds, three of them south of Trent, viz. Rushcliff, Bingham, and Newark hundreds, containing betwixt a third and fourth part of the county, and three north of Trent, viz. Bassetlaw, subdivided into North and South Clay, and Hatfield divisions, (which make it equal to three hundreds) Broxtow Hundred and Thurgarton a Lee. In the usual divisions of this shire Bassetlaw and Newark are equal to or set against the other four wapentakes, the town of Nottingham being left out.

SECT. III.—CLIMATE.

Being situated between fifty-two deg. fifty min. and fifty-three deg. thirty-four min. north latitude, it may be supposed to be later in its harvests than the more southern counties. There is however an exception to this with regard to oats and rye, which, in the warm gravels about Newark, are as early as in most counties, being often brought to Newark market before the first of August. The seed time and harvest may in general be stated as follows: Wheat seed time, from the latter end of September to the beginning of November, and often later; spring seed time, from the beginning of March to the beginning of May; turnips, from the middle of June to the latter end of July, hay harvest, from the middle of July to the middle of August; corn harvest, from the beginning of August to the latter end of September. The only particular circumstance that seems to deserve notice in the climate, is its dryness.* From my own observation, and that of many experienced persons I have consulted, I have reason to conclude, that much less rain falls in this county, than in the neighbouring ones to the west and north, which may perhaps be naturally accounted for by the clouds from the western ocean breaking upon the hills of Derbyshire and Yorkshire, and exhausting themselves before they reach Nottinghamshire; and even those from the German ocean may be supposed not unfrequently to skim over this more level country, and break first on the hills before mentioned: the greatest rains are observed to come with easterly winds. The drought of the summer 1793 was particulary experienced in this county.

SECT. IV.—SOIL AND SURFACE.

The surface of this county, except the level through which the Trent runs, is uneven, and may perhaps be said

---

* Vid. Appendix No. I. and compare with the Staffordshire Report.

to be hilly, though none of the hills rise to any considerable degree of elevation.

In point of soil this county may be divided into the three districts of 1. Sand or gravel. 2. Clay. 3. Limestone and coal land.

The sand or gravel may again be conveniently divided into, 1. The Forest country, or the borders of it. 2. The Trent Bank country. 3. The tongue of land east of Trent, running into Lincolnshire.

The forest district—consisting of the ancient forest, and the borders of it, of the same kind of soil, is in length (as may be seen by the map) about thirty miles, and in breadth from seven to ten, more or less in different places.

TRENT BANK LAND.

I consider as Trent bank land, the level ground accompanying the Trent, from its entrance into the county, down to, or a little below Sutton upon Trent, where the clay soil comes down to the river on the west side; and on the east, a poorer sand runs in a tongue-shape into Lincolnshire. I include in it, likewise, the level grounds running up the river Soar, from its junction with the Trent, up to Rempston,—as the townships of Ratcliff upon Soar, Kingston, Sutton Bonington, Normanton, and Stanford; and those lying on the back of them,—as East and West Leak, Cortlingstock, and Rempston, which, though on higher ground, are much lower than the Woulds, and of a good mixed loam, convertible, and equally fit for tillage or pasture; not let at less than twenty shillings an acre throughout, taking upland and meadow together; as well as the strip of higher land, on which are the townships of East Bridgeford, Kneeton, Flintham, and Stoke, which, though above the level of the rest, are of a mellow mixed soil, different from the clay of the vale of Belvoir, adjoining. This level is, in general, of a mellow soil or vegetable mould, on sand or gravel, though in some places these rise to the surface. It is of different breadths; in some places, not above a mile and a half; in others, three, four, and five miles wide; and is mostly inclosed.

The tongue of land east of Trent, is of a sandy soil, in some parts rather better than others, but in general very poor. A great part of it is taken up by low moors, much flooded by rains.

The clay country of Nottinghamshire may properly enough be divided into

I. *The clay north of Trent*, consisting of the north and south clay divisions, and the hundred of Thurgarton.

II. South of Trent, comprehending 1. The Vale of Belvoir. 2. The Nottinghamshire Woulds.

I must observe that the *clays north of Trent*, are in general not of so tenacious a nature, as in many counties, being more friable, from containing a portion of sand and falling more readily by the weather; particularly the red clay, of which there is a great deal in the country round Tuxford, and in the hundred of Thurgarton, which might perhaps be more properly called a clayey loam, and a blackish clay soil, commonly called a woodland soil, in which there is plainly a mixture of sand.

The Vale of Belvoir having no precise known boundaries, as soil with me is the chief distinction, I shall call by that name the country lying between the hills called the Nottinghamshire Woulds, and the strip of land running along the Trent on which stand the towns of East Bridgeford, Kneeton, Flintham, and Stoke; which, though not on the same level with the rest of the Trent bank land, is of a mellow mixed soil which will bear the same cultivation, quite different from what I term the Vale. The soil of this latter is generally a clay or loam.

THE NOTTINGHAMSHIRE WOULDS

Are a range of high bleak country; the townships are some open, some inclosed, as in Appendix, No. VII. The soil is generally a cold clay.

LIME AND COAL DISTRICTS.

The lime-stone and coal district may be defined to lie to the west of a line drawn from the river, at Shire-Oaks, pretty nearly south by west to the river Lene, near Woolaton and Radford, no lime being found east of the Lene. The lime-stone, which may be called a hungry lime-stone, rising up to the vegetable mould, commencing at Shire-Oaks, and beginning to abutt on the coal near Teversall, runs afterwards between it and the sand. The line of coal begins a little north of Teversall, runs about south and by west, to Drookhill; then south to Eastwood; afterwards about southeast, or a little more easterly to Bilborough, Woolaton, and the Lene. This line is scarce above a mile broad in this county, and above the coal is a cold blue or yellow clay. Between this and the sand of the forest, is the strip beforementioned of lime-stone.

SECT. V.—MINERALS AND FOSSILS.

Stone.—At Mansfield is got a very good yellowish freestone, for the purposes of building and paving, staddles, &c. and for cisterns and troughs, a coarser red kind. At Maplebeck, is a blueish stone for building, of which Newark bridge is built, which bleaches with the air to a tolerable white. At Beacon Hill, near Newark, is a blue stone for hearths, approaching to marble, which also burns to lime. At Linby is a coarse paving stone, much used at Nottingham.

Coals—are got in the line described in the Coal and Limestone district, and conveyed by the Erwash and Nottingham canals, as well as by land carriage. The price of them is of late greatly raised at the pits, owing probably, in great measure, to the enlarged demand occasioned by the extension of the navigation into Leicestershire, and the supply by the opening of new pits, not yet corresponding with it, but that evil is beginning to remedy itself, as many are now going to be worked, and the price is already considerably fallen at Nottingham.

Lime—is made of a weak *kind, for land* at Kirkby, Skegby, Mansfield Wood-House, and Warsop, of a better more soapy kind at Hucknall. On Beacon Hill near Newark, of a good kind, from a blue stone.

At Linby is exceeding good *lime for building.*

Gypsum or Plaster—is got of an excellent kind, on Beacon Hill near Newark, and much used for plaster floors.— A good deal of it is sent to London in lumps for the colourmen, and of the white, ground, in hogsheads, for other uses. At Red Hill, at the junction of Trent and Soar, is a fine plaster quarry, from which Mr. Pelham, of Brocklesby in Lincolnshire, now Lord Yarborough, had columns of twenty feet high, in three pieces, used in his mausoleum: Lord Scarsdale also used the same in his house at Kedleston.

Plaster is found also at G. Markham, and the Wheatleys, and in many other places, amongst the red loam; but I do not know of its being got for sale any where else than near Newark and at Red Hill.

Marl—Marling land is not used in this county, nor do I know of any marl pit opened; though there is reason to believe that there is much of it in the clay soil, as a red crumbling stone, and a blueish, are both found at Halam, Kirklington, Oxton, Gedling, and in many other places; effervesce strongly with the vitriolic acid, and if found in sufficient mass, there can be no doubt of the improvement of land from the use of it. The blue is in narrower veins than the red, and has a smell of sulphur when the acid makes it work.

N. B. I am since informed that Mr. Green of Bankwood, has lately found good marl on his farm at Saxendale in the Trent bank district, and is now beginning to lay it on his grass lands.

SECT. VI.—WATER.

This county may be said to be well watered for different purposes. The navigable river Trent enters the county near Thrumpton and runs through Nottinghamshire on both sides, till a little below N. Clifton, from whence to the northern point of the county it forms the boundary between it and Lincolnshire.

The Erwash forms the boundary between this county and Derbyshire for ten or twelve miles down to its junction with the Trent, a little below Thrumpton.

The Soar forms the boundary between Nottinghamshire and Leicestershire, for seven or eight miles above its junction with the Trent, a little above Thrumpton.

On the Forest side no less than five fine streams cross from east to west almost parallel to each other and afterwards run to the north forming the river Idle.

The Rainworth water runs from near Newsted Park, to Inkersall Dam and Rufford, and joins the Maun at Ollerton.

The Maun goes from Mansfield, by Clipston and Edwinstow, to Ollerton.

The Meaden, by Budby, and through Thoresby Park, joining the Maun near Perlthorp. From this junction the river is called the Idle.

The Wallin, through Welbeck Park, and after receiving the Poulter from Langwich and Cuckney, by Carberton, and through Clumber Park, into the Idle, near Elksley.

The Worksop river runs from Worksop by Scofton, Bilby, Blyth, Scrooby, into the Idle, at Bawtry. Two other rivers run southward.

The Lene, from Newsted Park, by Papplewick, Bulwell, Basford, and Lenton, into the Trent, by Nottingham Bridge. The Dover, or Dare Beck, from near Blidworth, by Oxton and Calverton, Eperston, and Lowdham, into the Trent, near Caythorp. In their course through the forest these rivers run mostly through boggy bottoms.

In the Clay District, N. of Trent, are the Dover Beck, the Greet, and many smaller nameless streams.

In the Vale of Belvoir, are the Devon, the Smite, and other smaller rills.

CHAPTER II.
State of Property.

SECTION I.
ESTATES AND THEIR MANAGEMENT.

ESTATES in this county are from about 12,000*l.* downwards to the smallest amount: nothing particular occurs in the management of them. Gentlemens' estates are, as in most other counties, under the care of stewards. Some considerable, as well inferior yeomen, occupy their own lands.

SECT. II.—TENURES.

Lands are holden as in most other counties, under a variety of tenures—freehold, copyhold and leasehold.

A good part of the small copyholds are borough English, *i. e.* descend to the younger son.

There are many leaseholds for three lives absolute (or freehold leases) holden under the archbishop of York, or the church of Southwell. Some pretty considerable estates formerly belonging to the priory of Thurgarton are holden by lease for years under Trinity College, Cambridge: The greatest part of the farms are let at will.

CHAPTER III.

Buildings..

**HOUSES OF PROPRIETORS, FARM HOUSES, COTTAGES,
&c.**

FEW counties, for their size, contain more seats of noblemen and gentlemen; a description of which, in an agricultural treatise, cannot be expected.

As many gentlemen keep a good deal of land in their own hands, they, as well as many substantial farmers who occupy their own, have made themselves extremely good farming coveniencies, so that in this respect, there may be said to be a very great improvement of late years. Several of the latter have indeed built themselves dwelling houses, much beyond the idea of farm houses in the last age.

Farm houses and offices are in general not very spacious, and in most parts of the country, *except in new inclosures,*[*] situate chiefly in the villages, and not contiguous to the land. Houses and barns generally, (except in the strip of country adjoining to Derbyshire, where there is plenty of stone, which is applied to that purpose) are of brick, and tiled, sometimes thatched. Poor cottages and barns, in the clay country, now and then of stud and mud; but new buildings of all sorts are universally of brick and tile.

---

[*] As a remarkable instance of this, Jonathan Ackom, Esq. of Wiseton, on the inclosure of Wiseton, Matterzey, Everton, Mission, and Scrooby, pursued the plan of placing new farms central to the respective grounds, and completed seven with large appurtenances, dove cotes, granaries, cow-houses, &c. The same has been done more or less in other new inclosures.

Ground floors are generally laid with stone or brick: chamber floors almost always with plaster, which is a great preventive against fire. Excellent plaster is got at Beacon Hill near Newark, and is run at nine-pence per square yard, or six-pence a strike. There is generally a good fold yard, and in the Clays North of Trent, very frequently a large good dove-cote. It is the custom of this country to put corn mostly into ricks, often set on stone staddles, or brick pillars, about three or four feet high, with stone caps; sometimes on brick hovels, open on one side, with pillars, or timber frames, about eight feet high, which leaves underneath a good shelter for cattle, or for carts and waggons. This custom, besides being thought to keep the grain sweeter and freer from vermin, is a great saving in the barn room expected in southern counties. It is of late come much into use with good farmers, in building stables or cow houses, to leave a space parted off three or four yards in width, behind or between two stables, into which the hay seeds fall from the back of the rack, and are saved for use, called a fother room; the rack is upright in the stable, and inclined on the back side. In improved cow-houses, the standing is made no longer than the cow herself. She stands on a kind of step, so that the dung falls down below her. Mr. Calvert of Darlton, has built some in this manner, but it is more used in Yorkshire.

Mr. Chambers of Tibshelf, in a letter to Sir Richard Sutton, describes a particular method of laying barn floors, as under:

" SIR,

" About twenty years ago I laid a barn floor with oak beams, fourteen inches square, and three inch oak plank, the plank was fourteen inches hollow from the ground, and the beams about two feet asunder; in two years after, some part of the plank broke down, without any other use than common thrashing upon; I examined the reason, and found the under side of the plank decayed by the damp rot, nearly through; upon which I had the floor taken up, and found all the planks in the same situation, and the beams almost totally perished; upon which I consulted a very experienced architect, who advised me to lay the next floor still higher than the former, and if possible to admit a circulation of air under the same, as the situation of the barn must be very subject to the damp rot. I relaid the floor with new beams and plank of the same thickness as the former; the beams were fixed upon brick pillars, fourteen inches high, so that the floor lay twenty-eight inches hollow; and under each door-sill was two grates, about one foot square each, that gave a current of air under the floor through the barn, and by the beams being laid upon supporters of brick, the whole floor was hollow except the nine inch pillars.

" The current of air was not through the middle of the floor, as the doors were more to one side than the other. In about two years the planks that were farthest from the passage of air fell down, all reduced to rotten wood, but about ¼ of an inch at the upperside; upon taking up the floor, I found the beams nearly reduced to rotten wood, except those that lay near the current of air, which were very sound, as was also the plank that lay over them in that situation.

" After these trials in the usual way of laying barn floors, I determined upon the following experiment:—to lay the next floor solid, in lime and sand mortar; upon which I removed every part of the former materials, and fixed fresh beams upon a spreading of mortar, at about six feet asunder, so as to suit the piecing of the planks to pin to; between each beam I filled the space with stones and thin mortar, that the whole was made solid with the upper sides of the beams; when this preparation was sufficiently dry, I culled the best of the remaining planks from each of the former floors, and before the workman laid down each plank, the space that I covered was spread with fine mortar, even upon the beams; then the plank was laid down and pinned; so that every hollow part, either in the beams, or decayed parts in the planks, was filled solid with the mortar. The floor has now laid about sixteen years without any amendment, except one of the planks being so weak in sound wood, that it started from the pins a year ago; after taking the same up and examining the underside, I found such of the plank that was sound when last laid down, was still perfectly so, and the rotten part was firmer and stronger than when laid down.

" I have mentioned the above circumstance to several workmen, &c. but few will follow the practice without peremptory orders, which I have followed up, with twenty different floors in sixteen years; some of oak, and some of deal planks, in different situations, and all appear as if they would stand for ages, unless they wear through from the upper side.

" About ten years ago I ordered a floor to be laid, with ash plank (in the method above) as oak could not be got without cutting down oak trees to considerable loss; the ash was fallen, sawed up, and laid down in the floor immediately, as the saw had left it; the joints did not open much before the last summer, upon which I had the floor taken up and relaid, as the plank must be well seasoned in nine years use; the under side of which was as sound as when first put down.

" There are other advantages by laying barn floors solid with lime mortar, as a barrier against rats and mice, worms, &c. Prevents the joints from wear, by quivering, &c. and also prevents any loss in the corn through the joints.

" I am pretty confident that the same method will hold good in laying ground floors with boards, in gentlemans' houses, that I have frequently seen them rot and eat through with worms in a few years. Both of these maladies the lime mortar will prevent, and will make the floor warmer also.

" It will hold good to cover the lath and pin, or nail, in the under side of slate and tiled roofs, by starting or rough drawing the same, so as to cover the lath, &c. close. My first observation of this simple method was about twenty years ago, being witness to the pulling down of an old timbered house, dated 1564; between each beam was pieces of ash wood, with split ash and hazel laths, and plastered on each side with lime or plaster mortar; where the plaster was free from cracks, the laths and the bark of them, was as sound and fresh as if they had not been cut down three months, but where the cracks had admitted the air, every part of the lath, &c. was reduced to nothing. I have lost no opportunity since my first observation, to examine the old ruins, and where I have found any old timber within those walls that had been run with thin mortar, the same has been sound and fresh, though the building has been in ruins for two centuries.

CHAPTER IV.

Mode of Occupation.

SECTION I.

SIZE OF FARMS—CHARACTER OF THE FARMERS.

THE farms may, in general, be said to be small, few exceeding 300l. per annum, and more being under 100l. a year than above that sum; many (especially in the Clays) as low as twenty pounds, or under.

The largest farms, as might be expected, are on the forest in the poorest lands, and which have been lately brought into cultivation. Many of the principal farmers carry on agriculture with great spirit, adopting the best practices of other countries; nor can it be said that the lesser farmers are backward in following good examples, of which they have seen the success. A very great difference may be seen from the face of husbandry twenty years ago.

SECT. II.

RENT IN MONEY—IN PERSONAL SERVICES.

Rents, as in other counties, are now universally in money. Some few boons (as they are provincially called) *i. e.* obligations to perform some carriage (chiefly of coals) for the landlords, are reserved, besides the rent, but to no great amount. On the other hand, it is not uncommon for farmers to give the carriage of coals to the cottagers who work with them as labourers, sometimes receiving their dung or ashes in return.

SECT. III.—TITHES.

Are in many places taken in kind, but are more frequently compounded for, at a much lower rate than they could be valued at by any surveyor. In the new inclosures, land has universally been given in lieu of tithes. As in other counties, there are modusses for different kinds of products, and some lands, which were anciently in the occupation of religious houses, are tithe free.

SECT. IV.—POOR'S RATES.

Poor's rates vary extremely in different parishes. It is impossible, as well as useless, in a work of this kind, to enter into the detail of them; I can only observe in general, that they do not run so high as in many counties, where manufactures have formerly flourished, which are now come to decay: but at the same time, it is a matter of concern to observe, that the manufactures, particularly that of stockings, whilst they increase the population, increase at the same time the burthen of the poor's rate on the occupiers of land; which may be ascribed to the lower manufacturers too frequently spending all their earnings, without looking forward to a time of old age and infirmity. The most obvious remedy for this evil, appears to be the extension of the Friendly Societies, which have already met with the encouragement of parliament; or the making of some more comprehensive provision by the legislature, on the same principle.

SECT. V.—LEASES.

The greater part of the lands in this county are, I believe, let to tenants at will, who in general do not feel themselves uneasy under their tenure, and frequently succeed to their farms from father to son for generations. Where there are leases, the covenants are the usual ones, as to repairs, not cross-cropping, &c. without any special provision that I am acquainted with.

SECT. VI.—EXPENCE AND PROFIT.

The expence and profit depend so much on the particular management of individuals, that I cannot pretend to enter into the subject. There is certainly room to make very fair profits on the farms in this county. Many fortunes have been made by farming, and the rents of many estates considerably raised without complaint or injury to the occupiers.

CHAPTER V.

Implements of Husbandry.

THE plough generally used in this county, is the Dutch swing plough, which is found to answer very well, their gate or bottom being from two to two and a half feet, with a pair of hales or handles, at a proper height to hold.

In the Vale of Belvoir the two-wheeled plough is used, which is made at Moor Green, near Nottingham. A one-wheeled plough is used near Nottingham, south of Trent, with two horses. A one-wheeled drill plough for turnips, is likewise made at Moor Green, which is much approved. The one-horse plough (on the recommendation of the late Charles Chaplin, Esq. of Tathwell in Lincolnshire) has been tried with success at Averham, Farnsfield, and Norwood Park. It appears perfectly sufficient for all the ploughings, (particularly on light land) except breaking up a swarth, and makes great dispatch in the sowing of turnips. At the latter place it was used this spring, on a strong red loam to sow barley and oats, and from this trial appears to do well even in strong land that has been previously worked pretty fine. In consequence of an experiment made at Norwood Park, Feb. 7th last, before some good judges, eight of these ploughs have been bespoke. They are made at Wragby in Lincolnshire, by Mr. Watkinson. Some of Mr. Cook's drill machines have been introduced, and approved. The drill part of the Moor Green plough may probably be easily adapted to this plough.*

The harrow is adapted to the soil; in the light soils, light and short tined; in strong and heavy soils, heavy and longer tined. Each harrow has its horse, three or four

---

* For the Cultivator or Scuffler used by Mr. Bower, vide Appendix, No. V.

are sometimes drawn without being, as in other places, connected together.

Mr. Jones of Arnold, has a curious threshing machine, which, with some kinds of grain, succeeds very well, but is expected to be brought to greater perfection.

Mr. Wright of Runby, has lately got a threshing machine, which answers his expectation compleatly, and he thinks no man who farms on a large scale, ought to be without one. They are made by Mess. Wigfall and Basmore, at Aldwark Mills, near Rotheram; the charge at present (for he hears they talk of advancing the price) thirty-five guineas for the machine only. The expence of putting up, together with timber for the stage and shed, if it is built with brick and tile, about 60l. more.

Waggons and Carts—The waggons of this county are of a middle size and height; the farmers have generally boards to fix on the top of their waggons, which they fasten with stays put into staples, and by these means they make them hold a larger quantity. For top loads or harvestings, they have raves or shelvings so called; these are in two parts, and put on at twice; they are very light, and when on the waggons are twenty-four feet long and eight feet wide, which gives plenty of room to load upon. They lie on the waggon as follows:

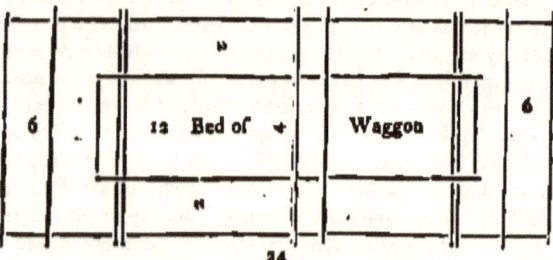

Waggons are very numerous; on small farms two, and upon larger three or four; carts being seldom used here in getting in the harvest. Carts here are made to tilt or slot, which makes them shoot their load at once.

CHAP. VI.

Inclosing—Fences—Gates.

INCLOSURE is going on rapidly in this county. There is seldom a session of parliament in which three or four bills are not passed for inclosing common fields.

The general fence used is whitethorn, planted in one or two rows, sometimes on the flat of the bank, sometimes on the front or slope of it. The posts or stoops generally of oak, with ash rails; though of late years, fir from gentlemen's plantations, particularly from Clumber and Rufford, has come much into use for rails. Gates are generally of oak, but willow with oak posts is found to make very durable gates, which have the advantage of lightness, and not damaging themselves by falling to.

In poor soils, which are particularly favourable to the growth of birch, birch has been found to answer very well for hedges, which may be pleached, and resist light stock, and also serves for light gates and rails.

The value of lands has been every where raised by inclosure, in a greater or less degree, in some very greatly.

Hedge-row Planting.—It is to be lamented, that in the new inclosures very little attention should have been paid to raising hedge-row timber, which is done at first with no more expence of fencing than the raising of the quick.

Whole tracts of country may be seen without a single tree growing up for farming use. This seems to arise from a mistaken notion of their being so prejudicial to the growth of hedges; which, as far as I have been able to observe, is not the case: they are likewise supposed to hurt the corn, which if it should be in some measure true with regard to ash particularly, it is easy to leave a headland of grass, or hedge green, as they are called in Hertfordshire; in some parts of which they are almost universal, and answer many good purposes. If any one will observe the difference of the number of trees, which may be raised by planting the hedge rows, instead of planting the corners of fields, as has been a fashion for some years past, they will be convinced that the former is the preferable method.

CHAPTER VII.

Employment of Land, Arable Land, Cultivation and Rotation of Crops.

SECTION I.

THE employment of land and cultivation depending in great measure upon the nature of the soil of the several districts, I apprehend I shall treat of it more distinctly by following it in order through each of them.

In the Forest District, the land being of a convertible nature, very little remains permanently in grass, except in the bottoms near rivers or brooks for meadow, and homesteds about farm houses for convenience. There was always about each forest village a small quantity of inclosed land in tillage or pasture, the rest lay open, common to the sheep and cattle of the inhabitants, and the king's deer.

Forest Breaks.—It has been, besides, an immemorial custom for the inhabitants of townships to take up breaks, or temporary inclosures, of more or less extent, perhaps from forty to two hundred and fifty acres, and keep them in tillage for five or six years. For this, the permission of the lord of the manor is necessary, and two verdurers must inspect, who report to the Lord Chief Justice in Eyre, that it is not to the prejudice of the King or subject. They are to see that the fences are not such as to exclude the deer.

Before the introduction of turnips and artificial grasses,

(generally called here, simply, seeds) which is scarce so old in the kingdom as the beginning of this century, and much later in this county, it was usual to get five crops running; oats or peas, barley, rye, oats, and lastly skegs,* then leave the land to recover itself as it could by rest. The introduction of turnips was of great improvement in insuring a good crop of barley after being fed off with sheep;† but still, till within these few years, it was not usual to lay down with seeds. At present, the culture of a break, well managed, may be stated to be—Break up for, 1. turnips, laying ten quarters of lime an acre; 2. barley; 3. rye, sometimes wheat; 4. oats, with seeds, *i. e.* white clover, and rye grass, which are mown for hay, and then thrown open. But the greatest improvement has been made in the forest lands permanently inclosed. ‡

Amongst these deserves to be named, in the first place, Clumber Park, belonging to his Grace the Duke of Newcastle, between ten and eleven miles round, and containing in the whole about 4000 acres, which may be said to be a new creation within these thirty years: at which time it was a black heath full of rabbits, having a narrow river running through it, with a small boggy close or two. But now, besides a magnificent mansion, and noble lake and river, with extensive plantations, which will be particularly noticed hereafter, above 2000 acres are brought into a regular and excellent course of tillage: maintaining at the same time between three and four thousand sheep, and are all in his Grace's own occupation.

The following courses and practices of husbandry, used

* Vide Appendix, No. II.

† I have met with a practice lately in Lincolnshire which may perhaps deserve imitation, viz. In order to make the turnip-tops of use, which are generally almost entirely wasted, to draw the fore teeth of culled ewes, which will then get fast on the tops without biting a root of the turnip.—R. Lowe.

‡ Vide Appendix, No. III. and No. IV.

in Clumber Park, were communicated to me by Mr. Birket, his Grace's farmer, a very active and intelligent person.

On the best Land.—First year, turnips; second, barley; third, clover; fourth, wheat; fifth, turnips; sixth, barley; seventh, seeds; which lie from five to six years.

On bad Land.—First, turnips; second, oats, with seeds which lie as before. The whins are stubbed constantly, to hinder this being obliged to break up sooner. He keeps a year's stock of dung before hand, and lays it on for turnips in autumn, ploughing directly. He harrows and gets out the twitch, (called in some countries couch-grass) as usual, in the spring. He lays two chalders, or eight quarters of lime an acre for turnips, but never after in that course. I shall defer saying any thing of the sheep, till I come to speak of that article in regard to the whole district.*

Newstead Park is now all brought into cultivation, divided into three farms.

Beskwood Park, containing 3700 acres, which before that time was little cultivated, except in breaks, was about nineteen years since taken on lease by a Mr. Barton, of Norfolk, who brought a colony of Norfolk labourers; but on some differences between them, they left him in a year or two. Mr. Barton is since dead, but the lease continues in his family; the land is now divided into eight farms. Many other tracts of ground have, in the course of twenty-five years, been inclosed from the forest and borders of it; some under acts of parliament, some by private proprietors without act, and some are now in agitation.†

---

* Vide account of the occupation of Clumber Park, Appendix, No. III.

† Vide List in Appendix, No. IV.

CULTIVATION AND EMPLOYMENT OF LAND.

The turnip husbandry prevails universally in such inclosures.* Sometimes, especially where the land lies in four

---

* The roota baga, or Swedish turnip, is now cultivated by a few farmers in this district. It appears to be superior to the common turnip in many respects, particularly in hardiness, as it stood the last severe winter without the least injury. It is eat with greediness by all animals, from the horse to the swine. Sheep prefer it to all others; but the material advantage that has been made of it is, the substituting it for corn in the food of draught horses; in which it has been found to answer the wish of every person who has yet tried it. The turnips are put into a tub or barrel, and cut small with an instrument like an hoe, with the blade put perpendicularly into the shaft; a man will cut in one hour, as much as six horses can eat in twenty-four. The tops and bottoms are previously cut off, and given to the pigs. Horses, that are hard-worked, look full as well when fed with this turnip and very little hay, as they formerly did when very high fed with corn. The Swedish turnip should be sowed early, from the 15th of May to the 10th of June.——R. LOWE.

The following information on the culture of the roota baga, has been given me by J. DAIKIN, Esq. of Nottingham:

" Mr. Daikin, about the § 10th of May, 1794, sowed about four acres with the seed of roota baga, about 2 lbs. per acre, on good sand land, worth twenty shillings an acre, manured as for turnips, and having been ploughed four or five times; the rest of the field, to the amount of nine acres in all, with common turnip and turnip-rooted cabbage, all broadcast. They were not transplanted, but hoed out nine inches asunder, at three hoeings, at seven shillings and six pence an acre: no other culture. In November began to use them for horses, giving at first clover and rye grass—hay, oats, and beans; but finding that the horses did well upon them, left off all corn, and continued them on hay, and the roots only: fifteen were thus fed for about two months, were constantly hard worked, and preserved themselves in very good condition. Mr. Daikin is so well convinced that in this application they were worth thirty pounds an acre, that he would, in future, if he could not get them otherwise, rather give that sum per acre for one or two acres, than not have them for this use. They lost their leaves entirely when the frost set in; but the roots were not the least affected, though the common turnips in the same field were totally destroyed. Passengers passing through the field cut holes in them, which did not let the frost injure them, nor were those hurt which were damaged by cattle biting them. Some came to the weight of

§ Mr. Daikin thinks that in general the roota baga should be sown about a month sooner than other turnips.

divisions only, the rotation of crops is, 1. turnips; 2. barley; 3. clover; 4. wheat; but more frequently it is let to remain longer under grass. It must be expected, that where every person follows his own ideas, there will be variations. The following improved courses have been communicated to me by able cultivators.

Mr. Wright of Ranby, whose mode of management has been adopted by several of his neighbours, 1. breaks up three year old lay for wheat on one ploughing; 2. turnips; 3. barley and seeds to lie three years. If a poor lay, prefers breaking: 1. for turnips; 2. barley with seeds, viz. rye grass, white clover, sometimes trefoil or hop clover, a little red clover, but does not like it, as hurting the rest. He pastures his first year's seeds with beasts, as sheep do harm the first year. The beasts are bought in at Michaelmas, kept the day on turnips, the night on straw. The seeds are ready for them by the time the turnips are done, and they go off fat in May and June. He weans his lambs about the fourth of June, and puts them on after the beasts.

16 lbs. and Mr. D. thinks the average of the crop 8 lbs. and much to exceed in tonnage per acre common turnips.

"Mr. Daikin gave them also to hogs, cattle, and sheep. They are excellent for hogs; and sheep being let into the field before the common turnips were destroyed, gave so decided a preference to the roota baga, that they would not settle on the common turnips, while the others were to be had.

"The method of giving them to horses, is to cut off the top-root, to wash them, and to cut them roughly with a perpendicular hoe, and then given directly, without keeping them to dry. The horses ate them with avidity, and seemed even to prefer them to corn. Their qualities appear to be singular, as they bind horses instead of relaxing them, as other roots do. One mare was kept intirely upon them and straw, worked every day, did well, and never looked better; this mare was more bound by them than the rest. They have a strong effect upon making the coats fine, and one or two, affected by the grease, were cured by them, as they act as a strong diuretic. In this mode of application, one acre maintained fifteen about two months, and Mr. Daikin is so well convinced of the utility of the plant, as well as many of his neighbours, that he intends, and they also, to increase the cultivation much.

"Mr. Daikin suspects there are two sorts of the roota baga, because some upon cutting are white within, but in general yellow; otherwise of the same external appearance. The yellow is the best."

After wintering his year old lamb hogs on turnips, he puts them on the second year's seeds; and they go off fat the latter end of May and beginning of June. On the third year's seeds he puts his ewes and lambs.

Mr. Bower of West Drayton, *on his best forest land*; 1. oats; 2. oats or (if the former crop is not shelled) sometimes wheat; 3. turnips, 4. barley with seeds for three years, viz. rye-grass, one bushel or two strikes, (the strike being the Winchester bushel); Dutch white clover, fourteen pounds; red clover, four pounds, or cow grass, *i. e.* red perennial clover, called sometimes red honey-suckle.

On worse Land.—Break up for, 1. oats; 2. turnips; 3. barley, with seeds for three years; one bushel of rye-grass, twelve pounds of Dutch clover: has a year's farm-yard manure before hand: chuses to lay it on at Michaelmas for turnips, if the land is nearly clean, and ploughs it in: then breaks with the drag harrow or cultivator, drawn by oxen, to get the twitch out; ploughs once in spring, then harrows with the common harrow, as often as necessary, and uses the cultivator:* does one and a half acres a day in ploughing, and with the cultivator seven acres a day or more. He sometimes goes out of his rotation, and sows blue and white peas in drill. Thinks the latter the best. He has a great opinion of buck wheat, as a cleaner of land. Finds cabbage a great impoverisher; sows trefoil, *i. e.* hop, or yellow flowered clover for sheep, not for a crop, as it does not rise high enough. Dutch clover grows as well, and is higher.

For Potatoes—trenches the swarth with a trench plough, then sets them in rows; in every third furrow about four inches asunder. Grows some of them every year; if two shillings a sack, would feed with them, but at the present price of three shillings and six pence, and four shillings, sends them to a market. He cuts them with two eyes to each set; sets nine sacks, and gets 115 per acre.

---

* For the description of the Cultivator, or Scuffler, as it is sometimes called, vide Appendix, No. V.

Rolling Seeds—is found a great improvement on light land, as I have myself experienced, using a stone roller.

Sowing Barley under Furrow—was tried last year by Mr. Jones of Arnold, who had a much better crop from it than when sowed at top; and found it stand the dry weather at least a fortnight longer. He ploughs his land as soon as possible after the turnips are off, beginning at one end of the field whilst the sheep are at the other. He thinks about the fifteenth of April the best time to sow his barley, which he does broad-cast, then ploughs it in. If he wishes to lay down with seeds, he then sows his seeds, and harrows. If he does not want seeds, he never harrows. Mr. Green of Bankwood, has tried the same experiment with the like success. For the kind of stock and manures, vide the general articles.

Potatoes are a good deal grown in the villages near Nottingham, (seldom above an acre together) for that market and home use.

Winter Tares—have been lately introduced. They are sown in September or October, November being rather too late, two strikes to the acre; and are an excellent food, cut green, for horses or other cattle.

PARTICULAR ARTICLES IN THIS DISTRICT.

Weld—or dyers weed, the *reseda luteola* of Linnæus, used for dying yellow, is grown a good deal about Scrooby, Ranskill, and Torworth; but the quantity varies much, according to the demand. It is sown with the barley and clover, half a peck to the acre; is pulled up from amongst the clover the next year, when the latter is coming in blossom, tied in bundles and dried. A good crop is half a ton an acre, a tolerable one six hundred weight. The price varies exceedingly, sometimes rising to 3s. a stone of fourteen pounds, or 24l. a ton; often 2s. 6d. and sometimes falling as low as 9d. a stone, or 5l. 6s. 8d. a ton.

Hops—are grown in this district: at Rufford about eighty acres, Ollerton thirty, Elksley thirty to forty. But as they are more the product of the clay, I must refer for the culture of them to that district.

Liquorice—was formerly much grown about Worksop, but is now entirely left off.

In the Trent Bank District.—The occupation is mixed of arable and grass, though more of the latter, especially contiguous to the river.

The Arable is generally calculated for the turnip husbandry, and kept in those courses, producing good crops of barley, and remarkable fine ones of oats, eight, and sometimes ten quarters an acre, particularly about Muskham and Balderton. They are so remarkably good, as to be distinguished by persons of knowledge from any other. Weight of the best, fourteen stone of fourteen pounds the sack; wheat is but eighteen. Oats are picked by hand by curious persons for seed. If the top one is a single oat, the rest on that stem will be so; the double ones are rejected. It is a strong instance of the improvement of husbandry, that about thirty years ago, the sand lands in Gressthorp Cromwell and Muskham fields, were not worth more than two shillings and six-pence an acre, covered with wild sorrel, and lea lay for six or seven years. The alteration is to be ascribed to turnips and clover.

The course of crops is often, 1. turnips; 2. barley; 3. seeds for one, two, three, or four years; 4. break up for wheat, sometimes for oats; sometimes 4. wheat; 5. oats; but not a general practice.

Joseph Sikes, Esq. at Balderton, near Newark, ploughs a lea on a sandy soil, and sows, 1. with fine Poland oats, (from six to seven strikes per acre) 2. rouncival peas; 3. wheat; 4. turnips; 5. barley; 6. oats and seeds, (viz.) 10lb. of Dutch clover, 8 lb. of red clover, 6lb. of trefoil, 6lb. of burnet grass, and 6 lb. of rib grass on every acre: the first spring after, stocks with sheep; gives the land a good top-dressing with rotten dung the next spring, in the month of

February, and stocks it with sheep and feeding heifers, for four or five succeeding years, and then breaks it up for the same rotation of crops. He prefers sowing seeds with oats to sowing them with barley, because the oats (Poland ones) being shorn (*i. e.* reaped) in this country, when there are fine crops of them, the seeds are not so apt to be smothered and damaged.

Mr. Sikes always found benefit to his wheat from mowing a second crop of clover, instead of earing it.

N.B. This is the Hertfordshire method.

Winter Tares are lately grown by some few persons, to cut for green fodder. Vide Ante Forest District.

Grass Lands are employed more for feeding than the dairy, except along the Soar, where, and in the towns on the south bank of the Trent, as far as Nottingham, viz. Thrumpton, Barton, Clifton, and Wilford, as well as at Attenborough, and Chilwell, on the opposite side of the river, are large dairies, milking from twenty to twenty-five cows, chiefly employed in making of cheese. The island between the towns of Averham, Kelham; Muskham, and Newark, is remarkably fine feeding land. Under the gravel here is found a clay, which is burnt into bricks; probably the same would be found in other places in this level. The beasts fed, are generally of the short-horned Lincolnshire and Holderness kind.

The Tongue of Land east of Trent, running into Lincolnshire, is of a sandy soil, in some parts rather better, but in general very poor. A great part of it is taken up by low moors, much flooded by rains. George Neville, Esq. of Thorney, has reclaimed a considerable part from the moor, and brought it into high cultivation; and has also, within twenty-five years, raised upwards of two hundred acres of very flourishing plantations. Some of the land, particularly where there is a thriving young plantation of seventy-five acres, laid out in quarters with ridings, appeared to me the worst I had ever seen, bearing little else naturally but the white lichen, or rein-deer moss. I cannot do so much justice to Mr. Neville's exertions, as by inserting in his own words, the account with which he was pleased himself to favour me, of his improvements and management. The remainder of this district being inconsiderable in extent, and poor in soil, does not afford any remarks on its agriculture.

Letter from GEORGE NEVILLE, Esq. to Sir RICHARD SUTTON, Bart.

Feb. 1, 1794.

" SIR,

" IN compliance with the desire you expressed when you took a view of my place with Mr. Lowe, that I would give you some more particular information concerning my improvements and management, I have to acquaint you, that I have taken from the best sort of common ling moorland, about 700 acres, and divided into five small farms, and built brick houses and necessary conveniencies, and tiled them, all upon plain useful plans. These farms are divided by quick fences, and many of the inclosures by birch; both grow extremely well. The birch fences, for the most part, are strong, and ready to plash in four years from planting: and by twigging them for besoms, pay something more than the expence at each plashing. They shoot very freely the first year after plashing, generally about the height of the hedge, 'and very soon make a strong fence. The best of the poor land is very thin skinned, not having above four inches of tolerable soil, before you come in general to a yellow, and too often a sharp white sand. The land is the best where the yellow prevails, being stronger, and not without a small mixture of clay in it. We plough thin, and keep all our manure as much upon the surface as possible. *Paring and burning* is never eligible, but always bad management, except where the land is first taken up from the rough, viz. when full of stalky old ling or gorse. Under these circumstances it cannot otherwise be brought into management in any moderate time, or at any moderate expence. The expence of paring, is from ten to twelve shillings per acre; burning, from four to six shillings; stubbing, more or less, according to the work. This management, or rather mode of taking up land, secures a good crop of turnips; and the layer for sheep always good. The course of tillage is for the most part, as in forest land; turnips, wheat, rye, peas, (grey rouncivals chiefly) barley on some parts; oats, lentils, and vetches, grow very well with good management, the whole art of which consists, (supposing the land judiciously worked and cropped) in keeping it fresh and fresh, viz. never to crop it much in tillage, or continue it long in seeds. As soon as the land can be spared, when the seeds begin to decline, it should be ploughed up again in the swarth for peas, oats, or skegs; then manure well for turnips; then lay down with proper seeds, whilst the land is in good heart, with a crop of barley, or winter corn, if the turnips can be eat off in time for the purpose. Rye grass has its use upon this land, mixt with white clover, trefoil, red clover, and perennial red clover, commonly known by the name of cow-grass, or in some other counties, red honey-suckle. Having no meadow land upon this sort of farm, I am reduced to mow the first year always, but never afterwards, pasturing it till again converted into tillage, with sheep. There is necessarily a variation from this system, according to circumstances and seasons. I raise heaven and earth, to make *manure:* but as very little clay, of any quality, is to be met with upon this sort of land, I have but few opportunities of making any large or sufficient quantity for the improvement of it. We use lime (Knottingley is by far the best) and to advantage, when mixed with good dung. A chaldron, with ten loads of manure, is more, in general, than we can get for an acre. Lime by itself on this sort of land is little used, except on fresh land. I mow ling, rushes from other lands, which I bring to bed with and tread down, and even pare waste land in lanes, to make *compost* with ditch stuff, fern, lime, rushes, dead gorse, &c.—By this management you may judge how I am straitened in the article of manure. Having two dove cotes, I use about ten chaldrons of *dove manure* every year, either on turnip land, or winter corn, with advantage. *Stickleback* manure exceeds all other that I have ever tried, as a present or lasting manure; but alas! I have not been able to get any for these last ten years: I used it then in tillage, and its effect is visible to this day. *Whale blubber* is to be procured from Hull, but the price asked is discouraging.

Of *sheep* I have two sorts, the one which I call my *upland* sort, of the Lincolnshire pasture breed. The hogs, run about three to the tod of twenty-eight pounds; the ewes, five or six.

As to my *lowland flock*, I began to breed from the Weighton, Yorkshire Limestone sheep, which was a small, compact, short wooled, hardy sort, and suited extremely well; but for many years I have crossed with the shortest wooled forest sort. I have now a flock of Scotch sheep, with black noses and feet. They are very hardy, and very round and compact in their make, and short wooled; these have been crossed by an excellent short-wool forester, and my lamb hogs are as beautiful as can be. My ewes of this sort, are this present season in lamb by a tup I bought with a flock of ewes, from Sutton Colefield. They promise to do very well upon my land, and I expect much from the crossing my breed. I never thought any thing about the rot, till the year before last, when I lost several, and again this year I had a loss amongst those I kept on; but it is very remarkable, that the Scotch ewes, which are now three years old, have escaped the rot altogether, though on the same sort of land, and when every body suffered greatly: these ewes turned out their lambs themselves, almost fat, giving as much milk as would rear two lambs each, and this without exception. I sold the wool of my short wool flock, at twenty-six shillings per tod, when I last sold, the year before last; at present I sell no wool. My short wool flock ewes tod, about ten to the tod. I have had wethers of this sort, which have come up to twenty five pounds per quarter. Those I kill for the house this winter, are from twelve to sixteen pounds, some eighteen or nineteen, one twenty pounds a quarter. I give great quantities occasionally, in the spring, of *skegs* in the corn to my lowland ewes before lambing, when, you can easily conceive upon such a sort of farm, I must be put to it frequently for support. I also chop the straw with the corn in it, and give it to them occasionally, and they are very fond of it. This year, having plenty of turnips upon my low farm, I have already begun to give some to a parcel of ewes which I expect to lamb about the middle of February. In this manner I support a flock of 120 ewes with their last year's lambs, on a farm which, when I began to inclose, had never been let at above three pounds a year.

" *Skegs* are remarkably good for horses, in the straw, or threshed, and in the straw remarkably so for cows. Straw beasts are also very fond of the straw. I can get taking up weak land from the swarth, from four to six quarters an acre. On the same land I should not have been able to get, perhaps, above three sacks of black, red Friesland, or what is called short small, or any other kind of corn. If land is in the state in which all our bad land will be, more or less, after the seeds are quite run out, and as is frequently the case, when you cannot conveniently put it again into tillage so soon as you could wish, from either the too great quantity you may have to improve, or other circumstances—if in this state, I say, it were to be taken up with turnips, the most eligible way, it would be often impossible to provide manure *upon the farm* for any quantity of land sufficient for the purpose, which makes *skegs* a very desirable crop, and which, if for that reason alone, I find a great acquisition, and feel a pleasure in having been the first to introduce them into this country.

"*Planting.* The scotch fir is the least nice of any of the sorts, and grows with me like a weed, and with as little trouble as gorse. The larch and Weymouth pine are equally hardy and prosperous upon all the land I plant, which is of the bad sort, and grow as fast and as well as upon the best soils I have ever seen. I plant birch in all my plantations as nurses to other plants, and as underwood; and in both cases they are very useful and advantageous, drawing 'up and keeping warm oaks, beech, ash, &c. which all grow as well as possible. *The beech* (under a favourable mode of planting) grows extremely kind and well in every respect; and I am now cutting ash poles, remarkably kind and good, by way of thinning my plantations of oak upon land never let before I began to farm and plant, at so much as one shilling per acre.

"As to the *uses of birch*. My birch is felling from November to the beginning of March, though the sooner the better, as it is very early in its sap in the spring, bleeding exceedingly if not cut before March. I first let the twigging to the besom makers at so much per bottle (or bundle), measuring four feet in the girth. The twigger lets the bottles lie till March, then takes them away, and stacks them like corn, and thatches them. They must, however, be tolerably dry before stacking, as otherwise they would be apt to heat and mould in the stack. The besom makers suit their own convenience as to the time of working up the twigs, generally beginning the latter end of the year, as soon as they become properly dry and seasoned for use. The making up is winter employment generally. I then cut out the shafts or staves, which I sell by the thousand or hundred. The tree thus dismembered I sell to the brushmakers, which is converted into brush heads, painters' brush handles, bannisters, spindles, distaffs, &c. and short pieces are worked up by clog makers and shoe heal cutters, &c. I sell these poles by the score, or by the groce of the articles they are converted to. In the last case they are cut up in the rough before they are carried from the woods,

The refuse is kidded up for the bakers and family use; the nogging ends unconverted, are brought into my own yard and burnt as coal, making the quickest and best burning fire possible, and the pleasantest, never sparkling or flying in the least. I used to raise a great quantity from seed. In this case I used to pare and burn a piece of my worst land, on which I sowed turnips, eat them off early, and ploughed the ground immediately, and harrowed it well. By this means the land was immediately in order for the seed, which I harrowed well in any time before the winter, which I prefer, as the plants will grow the sooner and make greater progress by coming out of the ground earlier than if this business was delayed till the spring. They will grow in great abundance sown broad-cast like corn, and be a nursery for years, leaving a sufficient quantity for a plantation at last. The seed may be easily taken from bearing trees, by cutting the branches before it is quite ripe in August, and may be threshed out with a flail as corn, as soon as the branches dry a little. By this means I can, in the course of a few hours, get as much seed as would sow an acre or two of land, at the rate of two strikes or bushels to the acre. I raise few in this way now, as I can furnish myself with seedling plants at one shilling per 1000 for taking up and planting the same at three shillings per ditto. These are taken up with small balls of earth, as are the *Scotch firs*, which I grow and plant out in this way, at four and five shillings, including the expence of taking up. These plants (whenever I plant them) do not seem to be sensible of their removal, and are carried with my own teams to the planting ground at no expence worth mentioning. The size of the Scotch fir then taken up, are from two to three and four feet high, and I do not lose one in a thousand, except by some accident. I have forgot to say that I make rails of my *birch*, tray, and gate bars, first shaving off the bark. The rails frequently are used to get up a second quick hedge on my bad land, so well do they wear; and I prefer them, for this purpose, to any oak or ash sapling rails whatever. They make very light and useful bars for inward gates for farmers, and with oak heads will last many years. I have corded a great deal of this wood, where it is not kind enough for riving for farming purposes, and it charrs very well.

"As to my *rein deer moss land*, as you termed it, I must say, bad and barren as it is naturally, it has for years worn its present coat, from the circumstance of its having been, till within these last twenty years, a turbary for generations; and I have known persons, who, getting turf upon it, have pared the same land three times over in the same day for fuel; so little was coal the fuel for poor cottagers, or even the better farmers, sixty years ago, where turf could be got. The turf and cutting it came to about fourteen pence per load in my memory.

"The *waste open moors* might and would be in their present state tolerably drained, if the waters in the Fosdike were not so kept up. My drainage is, for the most part, independent of this source of obstruction to all improvements; but I trust the late Lincoln navigation act will, in spite of every obstacle, be very useful to the drainage of the moor land. I suppose there may be 10,000 acres of open waste moorland between Newark and Bracebridge by Lincoln, including Swinderby, Thurlby, and other lordships in Lincolnshire and Nottinghamshire; the Scarles, Gerton, Spaldforth, Wigsley, Harby, Clifton, Swinethorp, Saxelby, Broadholm, Thorney, &c.

I am, dear Sir,

Your obedient humble servant,

G. NEVILLE.

"P. S. Since I had the pleasure of seeing you here, I have sold ash hop-poles (yesterday) from a plantation on the bad land, at forty shillings per hundred, which I only mention to shew that ash poles may be grown to advantage upon it. The birch hop-poles, very long and kind, I have sold at thirty shillings per hundred. I am now planting about six or eight acres in this sort of lands, in alternate rows of oak, ash, beech, and birch, which I prefer to alternate plants in the same row, planting them in quincunx order, so that the rows are lost as the plants grow up, and they become promiscuous, and regularly set at what distance you please."

In the Clay District, North of Trent,—there is a great intermixture of open field and inclosed townships; but more of the former, as may be seen by No. VI. in Appendix.

In the open field, the common course of husbandry is pursued as 1. fallow; 2. wheat or barley; 3. beans, pease, or both mixed. The latter crop is very common in this country. The reason given for it is its smothering the weeds; but I have always observed the crops to be very foul.

Folding is little used. Few farmers, indeed, have stock enough of sheep to do it with any effect.

In some places, of late years, clover has been sown with the barley, and mown the third year, instead of the bean crop, which, in lands that have been long in tillage, is often very poor. The old way, in Oxton fields, was the usual one of two crops and a fallow, there being only three fields. In consequence of the act for cultivation of common fields, of 1773, they have now sown broad or red clover with their wheat or barley, (except a few who chuse to have their old crop of pease and beans the next year) they mow the clover the second year, and then stock it with three horses to two acres; or else two cows, or six calves, or three sturks to an acre, and then fallow, except a few persons who let the clover lie another year, and then sow it with wheat. They find this answer so well, that they intend to divide one field into two, so as to have four fields. One barley, one clover, one wheat, and one fallow. In inclosed lands, part is kept as arable, part pasture. Fallows are still retained, sometimes in the old course of the common field. These different courses have been practised: 1. fallow; 2. beans; 3. barley; 4. artificial grasses two or three years; then, 5. wheat; or 4. red clover for one year only, 5. wheat: or, 1. fallow, with dung; 2. barley, with seeds; 3. 4. and 5. years, pasture; 6. break up and sow beans, with pease, or rouncival pease alone; 7. wheat. The following course has also been tried with success: 1. fallow; 2. barley or wheat; 3. beans; 4. red clover; 5. wheat. The clover crop sown with the beans, and the wheat crop, have both been remarkably good—at Norwood Park, by Sir Richard Sutton; by Mr. Musgrave at Halam; and at Red Hill, by Mr. Cook.

Mr. Turnell, of Stokeham, in inclosed lands, tried crop and fallow alternately, for twelve years; but did not find it answer. Potatoes are grown, but no where in large quantities, seldom above a land or two together. Most cottagers have a plot of them, which is of great use. The land is generally too strong for turnips to be fed off. Rape is sometimes sown instead of them, for sheep feed; sometimes for a crop, yielding half a last, or five quarters, often four; medium price, twenty-five pounds a last; sometimes thirty three pounds, sometimes fifteen pounds only. Rape has lately been employed with success to feed beasts, giving it to them mowed under sheds in the winter, and leaving the stalks to afford sprouts for sheep feed in the spring.—Query, if borecole might not afford a greater produce? Scarce any oats are grown.

Winter tares—begin to be introduced for cutting green.

Hops.—Are a considerable article of produce in this district, principally in the part about Retford, and some about Southwell and its neighbourhood. They are generally known among traders, by the name of North Clay hops; they are much stronger than the Kentish, going almost as far again in use; but those who are accustomed to the latter, object to their flavour as rank. The quantity grown is fluctuating, some yards being laid down every year, and others taken up. It is supposed there are not so many now grown as thirty years ago, but that the culture has increased within the last ten years. Mr. Bower, to whom I am indebted for the accurate account of the management of them in the following letter, supposes the number of acres now so employed in the whole county 1100: others carry them as high as 1400.

West Drayton, January, 7, 1794.

" SIR,

"When I had the pleasure of spending a few hours with you at Clumber, you desired me to procure you an account of the number of acres planted with hops in this county, and likewise the expence of their cultivation. According to the best information I can get, there is about 1100 acres of hop ground. It consists of different kinds of soil, but chiefly strong clay, and bog or black earth. This, (the Eastern) part of the county is strong clay: The plantations here lie in vallies and wet lands for the most part, not very valuable for other purposes. The common price of first taking up, or converting grass land into hop ground, is fifty-shillings per acre; exclusive of sets, planting, and draining, which cost about as much more. Poling new ground is the heaviest expence; it will (if new poles are used) cost at least twenty-five pounds per acre. The best managers here set but two poles at each hill, (and where the bind is strong) but two binds upon each pole.

"The common price of what is provisionally called, 'looking after an acre of hop ground,' is from forty to forty-five shillings: this work only consists of digging, picking, cutting, poling, twigging, once hilling and hoeing, and poles stacking. Good managers add at least two hoeings more, which cost half a crown an acre each. Then there is is the draining every other, or every third year, with fresh earth getting, catch poles sharping, carrying in, and setting, with many other little works, which are for the most part here done by the day: The men's wages are fourteen pence from Michaelmas to May-day, and eighteen pence from thence to Michaelmas again. It is to be observed, that the shopman has the grass that grows on the drains, with the broken pole ends, and often the binds. Upon taking an average of the expence of labourage of my hop grounds for four years, I find it cost me four pounds per acre; for the working part only. The poles, manure, rent, and tithe, about nine pounds ten ten shillings per acre; which brings a certain expence of thirteen pounds ten shillings per acre; if there is not a hop grown.

" The crops in this country, in the best years, are very small, compared with the Kentish plantations, and do not in the very best of years average eight hundred weight per acre; owing I apprehend more to the number of small planters, who have neither knowledge or purse necessary to this intricate and expensive culture, than the badness of the land. With respect to manures, the greater difficulty in the management of this plant, is to procure hops of a large size. The bind is easily forced by the use of rags, but they, if not properly used, will make the hops small. I apprehend the best way of using them, to prevent that evil, is not to lay more than eight hundred weight per acre, mixed with three or four cart loads of good virgin earth, of a light black soil, or strong land. This composition to be made at least six months before laying on, and turned two or three times, then laid upon the hills after the tying before Midsummer.

" I have used the scrapings and parings of oil leather, instead of rags mixed as above, and have found them excellent manure.

" Malt culm I have likewise used, and think it a good manure, and particularly so, where land is subject to the small snail or slug, which eat the young bind on its first appearance; for it sticks so fast to their slimy bodies, that they cannot creep over it to the bind. Where land is subject to grow small hops, I am well satisfied the best method, (where it can be got even at at a very high rate) is to dig in, in winter, from twenty-five to thirty cart loads of good dung, and if it is not quite so rotten as to cut with a spade, I think it is the better for strong land. If this method, with good drainage, and keeping the land clean from all kinds of weeds, has not the effect of making the hops a good size; I should apprehend the land is either not congenial to the growth of this plant, or otherwise has been planted too long, and wants laying down to rest.

" The worst evil that attends the culture of hops, is the smitt, which nobody seems properly to understand, and for which no effectual remedy has ever yet been found out; neither do I think it would tend to the profit of the planters, whatever it might do to the public at large, to have such a remedy, as either more than half the land must be laid down, or hops would want a market.

" I am afraid these few observations, from their being well known to every observant and practical hop-planter, will not be worth your acceptance; but if I can be of service to you in any future inquiries upon this, or any other business, I shall always be happy to give such assistance as my small experience and abilities enable me to give, and am with the greatest respect,

SIR,

Your most obedient

humble Servant,

MARTIN BOWER.

To Sir Richard Sutton, Bart.

LETTER TO SIR RICHARD SUTTON, BART.

West Drayton, Feb. 1, 1794.

" SIR,

"I am favoured with yours, and in answer inform you, that the hop plantations beginning at the north end of the county are chiefly situated in the following parishes:

Barnby-Moor,	East Drayton,	Bothamsell,
Tiln,	East Markham,	Walesby,
Lound,	West Markham,	Willoughby,
Welham,	Milnton,	Kirton
Ordsall,	Tuxford,	Boughton,
Eaton,	Egmanton,	Ollerton,
Gamston,	Bevercoats,	Edwinstow,
Headon cum Upton,	Haughton,	Wellow,
Askham,	Elksley,	Eakring,
Woodcoats,	Southwell,	South Markham,
Rufford,	Halam,	Laxton.

"There may be some single plantations in some other townships; but these are the principal."

Pasture.—Most farmers have some dairying keeping, from five or six to ten or twelve cows, in general perhaps eight cows to 200 acres, chiefly a woodland breed; but it is not their principal object, except about Fledborough, and from thence close along the Trent, down to Gainsborough, where the numbers kept (especially at Fledborough) may run as high as thirty.

The grass grounds along the Trent, in the open townships, are generally shut up at Lady-day, some part opened for stinted pastures at old May-day, some kept for hay, and all commoned from old Lammas. A good many young cattle are reared. In some places, particularly in the north clay, there is more feeding.

Mr. Turnell, of Stokeham, has a particular method of feeding which deserves to be mentioned. He feeds about eighty head of beasts a year, from twelve to twenty of his own breed. He always buys in at spring, takes care to have twenty to thirty fresh beasts, or incalved heifers, mostly of both sorts, that are sure to go off before the middle of July; when he lays in, *i. e.* shuts up about forty acres, to finish or make fat all his beasts that went lean at spring.* These go off in succession till Christmas. The land laid in (where good) will be very often ready to pasture again in six weeks. Mr. Turnell gave me the following instance of the difference *of sand and clay land, in bringing the sheep forward.* He feeds the same breed as Mr. Wright. At one year old, Mr. Wright's were worth 1l. 19s. and Mr. Turnell's of the same age and breed, not more than 1l. They were compared in April and May. Mr. Turnell's, he owns indeed, had not been fairly kept. In the February after, sixty out of ninety of Mr. Turnell's were sold at one year and ten months old, for 2l. 6s. Clay ground will not bring them so forward, they must be kept another year. *To shew the difference of the value arising from the breed,* Mr. Turnell told me, some of the same breed as his, kept by Mr. Wright's brother on good clay land by the Trent side, were sold for 2l. a piece, besides the wool, at one year and a half old. Two amongst

---

* This practice agrees with that of fogging, mentioned in the Cardiganshire Report. Sir R. Sutton has tried it with success for keeping his stock in winter. Many years ago being in the practice of buying in Highland Scotch bullocks about October, which he began to kill for his family about that time twelvemonth, he used to give them the first winter straw and sometimes a little hay, and did not find them go on to his mind. It seemed indeed to be the opinion of the neighbourhood that they would not feed kindly. His bailiff falling into conversation with a Scotch drover, and telling him how he managed them, the drover told him he would spoil them if he gave them any dry meat at all. It would (as he expressed himself) dry them up. Since that they have never had any thing the first year but such fog as they could pick up, and have had a little hay to finish them, only when the weather has been very hard frost or snow.

them, got casually by a bad tup, and which had always been with the others, were worth no more than one guinea.

In the Vale of Belvoir district—The country is part open, and part inclosed; as in Appendix, No. VII. In the open fields, the course of husbandry is generally, 1. wheat or barley; 2. beans; 3. fallow. In Elston* are four fields, as 1. wheat; 1. barley; 1. bean; 1. fallow. In the inclosures there is almost universally a mixture of arable and pasture, and a little dairying.

Clean fallows are generally made in rotation, as in the open fields. Sometimes red clover is sown with barley, and broken up instead of a fallow. Sometimes white clover, rye-grass, and rib-grass, or narrow leaved plantain, is sowed with the barley, and let lie three years.

Mr. Pocklington, of Kinnoulton, an active and spirited farmer, has had sixteen crops running, (including clover) without a fallow. He is careful always to hoe out weeds early in spring, as soon as they appear. At harvest he mows his stubbles directly; sets his crop in lines, and, if possible, ploughs the intervals before it is carried. He then ploughs it in to make the weeds vegetate; and as soon as that happens, if dry weather, ploughs again as soon as possible. He lets no wet stand on any account; drawing it off with the spade, or otherwise. He allows, that in wet lands fallows may be required oftener. The land in general is too strong for turnips. Rape is sometimes got instead of them, but rarely.

In the Nottinghamshire Woulds—The arable is occupied in the common course of two crops and a fallow. Clipston has four fields; the inclosed part is chiefly poor land; a little is ploughed, very few turnips, some clover. The Woulds, properly called, or wastes, in the open parishes, are stinted pasture for young beasts and horses. Few sheep.

* Elston is since inclosed.

In the Lime and Coal District.—The farms in general are small, the occupation mixed, but much arable, as the land will not either on the lime or coal, lie in pasture longer than the artificial grasses will last.

The course of crops are,

On Limestone—1. fallow; 2. barley; 3. grass seeds: or 1. turnips; 2. barley; 3. seeds, for two years, seldom more; 4. peas or oats, &c.

On Coal Lands—1. fallow; 2. wheat, sown at Michaelmas, and seeds sown on it at spring; 3. seeds pastured or mown, seldom let to lie above one year, sometimes two or three:—or, 1. seeds broke up for wheat; 2. oats; sometimes rouncival peas, then fallow again: or, where the land suits it, 1. turnips; 2. barley or oats, with seeds, &c.

This land is very subject to throw out the wheat.

Mr. Chambers of Tibshelf, in Derbyshire, who occupies similar lands abutting on Notts, and whose practice is looked up to and followed by several intelligent farmers in Notts, generally pastures the first year, dungs in the next autumn, mows the next year, then breaks up at Michaelmas or spring, for wheat or oats.

SECT. II.—CROPS COMMONLY CULTIVATED.

The crops usually cultivated in this county are the common ones, of wheat, rye, barley, oats, beans and pease.

The kinds of *wheat* commonly sown are, the red lammas, and white chaffed, or Kentish; two strikes, or Winchester bushels, sown to the acre. It is very difficult to form an average of the product of this, as well as other grain, from the great difference of soil and management.

The crops in common fields, may be said to be from two to three quarters, or more. In inclosures from two and a half to four.

A good deal of bearded wheat, here called Yeograve wheat, used to be sown particularly in the clay open fields, but is now much left off. It is a stronger stemmed hardier wheat, but coarser grained.

Rye is but little sown, being scarce at all used for bread. Chiefly in the Trent bank land, about Markham, &c. and on the forest. There are two kinds; a black, and a white or silvery sort. This latter has been sown in the spring with success.—Two strikes sown, crop from three quarters to four.

Barley is much sown—No particular kind is distinguished. The Fulham rathripe, or early barley, has been tried in a few places; but though ten days or a fortnight earlier, has not answered in point of quantity of produce, or boldness of grain, so as to recommend the practice.

Four strikes are sown, produce per acre, from three to six and seven quarters.

Oats of various kinds are sown.—In the Trent bank chiefly the Poland; in other parts the Friesland Holland oats, brown oats, black oats in cold lands, and a few Tartarian, which will grow on worse lands than others, but are later in ripening, and coarser grained.

Four strikes are sown; crop from four to seven; sometimes as high as ten quarters.

Skegs, a species of oats, are I believe confined to this country. Eight strikes are sown; and yield a crop double that of other oats, in quantity; but not more than equal in weight.—They will grow in the poorest land, and are reckoned very sweet food. They are seldom brought to market; but esteemed by farmers for their own use, and are often given in the straw.

Beans are almost all of the small horse bean kind. Four strikes sown, crop three to four and five quarters.

Pease.—The common blue pea is sown in the poorer lands; in some stronger, the rounceval; in some good Trent bank land, grey or white; four strikes are sown, crop from four to five and six quarters.

Buck Wheat is scarce enough sown to be called a crop of this country. It is found good food for pigs and poultry; and Mr. Wright, of Ranby, has found it useful for fodder for his team, but is tired of the system; as he has found it at length to be a hazardous crop, being subject to be much injured by slight summer frosts, which go near to destroy the crop, if they happen when the plant is in a particular state. The quantity he used to sow was six pecks an acre; average crop, from three to four quarters; but has been told of seven quarters, on rich soil.

CHAPTER VIII.

Grass Natural and Artificial.

SECTION I.

THE Banks of the Trent and of the Soar produce very good natural grass, which is applied to the use of dairying, feeding and mowing, as has been mentioned more particularly in speaking of the employment of land in the several districts.

The kinds of artificial grasses sown are—For a single crop to break up again for wheat, red or broad clover, (trifolium prateuse) to lie two or three years or more in pasture, white clover (trifolium repens) trefoil (medicago lupulina) ray grass (lolium perenne and rib grass (plantago lanceolata) the usual quantities of each of which has been mentioned under the head of cultivation, sometimes a small quantity of red clover is mixed.*

Burnet (poterium sanguisorba) grows naturally in plenty in the Trent meadows, but is not that I know of sowed.

Sainfoin (hedysarum onobrychis) has been tried on sand and gravel, and on a red loam with skerry stone underneath,

---

* Vide Mr. Calvert's letter in Appendix No. IX. Dr. Coke of Brookhill, in the lime and coal district says, the grasses which are cultivated for pasture are red and white clover, trefoil, rib-grass, ray-grass, and those seeds which are the natural production of the soil, and which consist of the anthoxanthum vernale, and several of the festucas, some of the aira and avena of the Linnean system; the former of these grasses is a most useful addition to these pastures, from its early appearance in the spring, and from the sweetness which it affords to the hay, of which it forms a part. The festuca fluitans, which is mentioned by Stillingfleet, in his Botanical Essays, is an inhabitant of the swampy parts of this neighbourhood, and is always sought for by cattle with the greatest avidity.

but without success, the soil does not seem to agree with it; it prospers best on a creachy soil, or thin skinned limestone.

Lucerne (medicago sativa) has lately been tried by some gentlemen in small quantities, and promises very well.

The method I pursued last year in laying my land down with grass seeds was attended with great success. Instead of sowing my land with barley and seeds after turnips, as usual, after one ploughing, which was about the last week in April, I sowed half a peck of rape seed, mixed with it one stone of Dutch clover, and one of trefoil; the rape got sufficiently high to shade the young plants from the sun. About the second week in July were turned upon ten acres, one hundred and thirty wethers and ewes, part of which went fat to Rotherham market, about the 11th of September; and the remainder I should have sent the latter end of the month, had not mutton fallen in price.

SECT. II.—HAY HARVEST.

Is in this county generally in July. Much cannot be said in commendation of the method of making meadow hay, compared with that of the south of England. It is very little shaken, and frequently not tedded; lying for some days in the swathe, and only turned. When once got into cock it is seldom spread out again, unless much wet comes; but is generally thrown into wind rows, or large swathes, before it is loaded. However, an improvement may be seen every day in making the hay with more care.

The fear of breaking off the heads is perhaps a good reason for not giving clover hay too much shaking.

CHAPTER IX.

Gardens, Orchards, and Nurseries.

THERE are in several parts of the county considerable market gardens and nursery grounds, particularly about Newark; amongst which are particularly eminent those of Mr. Ordoyno, who has been very industrious and expert, as well in raising exotics as native plants. In the Clay district are many orchards of apples and pears. Among the most considerable are those about the villages of Halam and Edingley, about Southwell. The making of cyder or perry is not practised; though the soil being very similar to that about Upton in Worcestershire, and Ross and Ledbury in Herefordshire; viz. a red marly loam with blue veins seems to promise success, and the trees may be observed to grow remarkably well and strait. One reason why this may not have been thought of, is the very ready sale of the fruit at Mansfield Market, for the supply of the Peake country, in Derbyshire, where the climate does not admit of orchards to any advantage.

CHAPTER X.

Woods and Plantations.

THERE are many woods and plantations in this county.

In the forest district.—The principal remains of the ancient forest woods are, the Hays of Birkland and Bilhagh, being an open wood of large old oaks, most of them decaying, or stag headed, and without underwood, except some birch in one part; it extends about three miles in length, and one mile and a half in breadth. By a survey taken for the crown, in 1790, there were found in both together, ten thousand one hundred and seventeen trees, valued at 17,142l. The land on which they grow is one thousand four hundred and eighty-seven acres, and is supposed would be worth, when cleared of wood, and inclosed—Birkland, eight shillings, and Bilhagh, twelve shillings an acre.

In a survey of 1609, were found 21,009 trees in Birkland, and 28,900 in Bilhagh; the trees in general were then past maturity. By a survey in 1686, there were 12,516 trees in Birkland, and 923 hollow and decayed ones. In Bilhagh 21,080, and 2797 hollow trees.

By a survey in 1790, there were in Birkland and Bilhagh together, 10,117 trees, at that time estimated at 17,147l. 15s. 4d. In the year 1609, there were in Birkland and Bilhagh, 49,909 trees; so that in seventy-seven years, to 1686, had been cut down 12,593 trees.

There are now and then opportunities of knowing the ages of oaks almost to a certainty. In cutting down some trees in Birkland, letters have been found cut or stamped in the body of the tree, marking the king's reign, several of which I have in my possession. One piece of wood marked J. R. (James Rex) was given me by the woodman who cut the tree down in the year 1736. He said that the letters appeared to be a little above a foot within the tree, and about one foot from the centre; so that this oak must have been near six feet in circumference when the letters were cut. A tree of that size is judged to be about one hundred and twenty years growth. If we suppose the letters to be cut about the middle of the reign of James the First, it is 172 years to the year 1736, which added to 120, makes the tree 292 years old when it was cut down. The woodman likewise says, that the tree was perfectly sound, and had not arrived to its highest perfection. It was about twelve feet in circumference. I have been told that J^n. R. (John Rex) have been found cut in some of the oaks. One piece said to be marked with John Rex, and a crown, I have in my possession; but it is not sufficiently made out to be inserted here as a fact, though the person from whom I had it assures me, from his having seen others more perfect, that it is marked with with Jo^n Rex. Others have had C. R. and several have been marked with W. M. (William and Mary) with a crown.*

A part of this wood has been taken by grant into Thoresby Park.

Harlow Wood, Thieves Wood, and the scattered remains of Mansfield Woods, are of small extent, and inferior size of timber.

In Clumber Park are the remains of two woods of venerable old oaks, called Clumber Wood, and Hardwick

---

* Descriptions and Sketches of some remarkable oaks in the park at Welbeck, p. 16, by Hayman, Rooke, Esq.

Wood. Since they have been shut in from cattle, the young trees are springing up surprisingly, from the acorns. For the woods and plantations in this district, vide general list, and subsequent ones of particulars.

Plantations.—The spirit of planting has prevailed much in this district since about forty years. Unfortunately, the first plantations were chiefly of firs, whether from the desire of making an early appearance, or from the notion that forest trees were not easy to rear in this soil. It has, however, been found since, that trees of all kinds, well planted and properly sheltered, succeed very well.

His Grace the Duke of Newcastle's plantations within Clumber Park, amount to one thousand eight hundred and forty-eight acres.

In the extensive inclosures made by his Grace in Elksley, Bothamsell, &c. the quick hedges, which are remarkably fine, were raised with posts and rails, the thinnings of these plantations. I was assured, some years since, that sixty miles running measure had been done in this manner; and by this time it must amount to double that number.

On the METHOD OF PLANTING, *as practised at Clumber, by the Duke of Newcastle, under the management of T. Marson. Communicated by Mr. Marson, the manager of his Grace's improvements.*

THE method in practice here, for the management of a sandy soil covered with heath, whins, or furzes, &c. and which is intended for planting, is as follows:

The ground is first cleared by stubbing and burning the heath, &c. (if found too strong to be ploughed in); it is then sown with one or two successive crops of spring corn, as oats and barley, and afterwards with turnip seed, the ground being first thoroughly cleared of twitch and other weeds. The turnips are fed off early with sheep, and the ground is immediately double-trench ploughed from twelve to sixteen inches deep, if the soil admit of it. We then proceed to sow it with acorns, ash keys, and hawthorn berries, and to harrow them twice over. The same ground is afterwards planted with oaks, ash, beech, elms, birch, larches, and other sorts, which you may occasionally have, from six to eighteen inches in height, but not to exceed that size. Spanish chesnuts are then put into holes made with a dibble, from three to four inches in depth, and covered over.

In general we fill our plantation with a various assortment of American plants; as firs, pines, cedars, &c. besides Scotch firs and birch. These are not only a shelter for the young forest trees, but have a pleasing effect for fifteen or twenty years, for their permanence of verdure, and variety of their foliage. They are then taken out to give room for the oak, Spanish chesnut, &c. This is the last process of our plantation.

For a more particular account of the rearing and management of oaks, on a sandy light soil, I beg leave to refer to the communications of the very ingenious Mr. Speechly, of Welbeck, which are inserted in Evelyn's Sylvia, published by Hunter, page 90, first part, (in the note) which explain every thing that can be practically useful on this subject.

EXPLANATORY OBSERVATIONS ON THE ABOVE SYSTEM.

The quantities required for one acre, are nearly as follows:

Plants of different sorts, from six to eighteen inches, about two thousand.

Acorns, from four to six strikes.

Ash keys, four strikes.

Hawthorn berries, one ditto.

Spanish chesnuts, one ditto.

The ash keys and hawthorn berries, are to be buried one year in beds or pots of sand, before they are sown in the plantations.

Six strikes of fine large acorns are supposed sufficient to sow one acre of ground, at one foot asunder.

The plants should be heathful, strait, and firmly set; not drawn up too luxuriantly, but raised in a soil of a similar quality and temperature to that wherein they are intended to be planted. This is a very necessary caution, as they will hence be much more likely to flourish, than if they should have been raised from a richer soil or warmer situation. They should also be always transplanted from the seed beds into fresh ones in the nursery, from four to six inches asunder.

If the heath can be ploughed in, the best time will be in autumn, as it will then have the benefit of the winter fallow. It may be cross-ploughed in the spring; and after having been repeatedly harrowed, may be burnt, and then sown.

With respect to the season for planting, I prefer the autumn for a dry sandy soil, and the spring for meadow or low ground; and also for sowing the acorns, berries, keys, &c. At that time of the year they are not so liable to accidents by vermin or frosts, as when they are sown earlier. When, however, planting and sowing is intended to take place in the same piece, both must be completed at the same season, that is, early in the spring. Precautions should be used to guard them from crows, hares, rabbits, and other noxious animals, till the chesnuts, acorns, and plants, are sufficiently grown not to be exposed to such casualties.

If the land be hot and dry, you may sow some rye or oats, with acorns, and other seeds, but not so thick as to produce a regular crop.

I am never under any apprehension of planting too thick, for many reasons which it is needless to enumerate. When the season proves favourable, and there are more live plants upon the ground than are necessary, they are thinned at a proper period, and made use of in the next year's plantation. This mode permits me to appropriate the nursery to other uses.

About fifteen years after planting, the trees may serve for a variety of purposes; such as posts, rails, pails; punchwood, for the colleries; cordwood, charcoal, hop poles, brush heads, birch brooms, joists, rafters, &c. &c. This observation is to be confined only to such trees as are cut down in the course of the second or third thinning.

The plantations are in general inclosed with quick fences. These, independently of their ornament, not only secure the young trees from being injured by cattle, but prevent the sheep from depositing their dung, which in that case, as a manure would be of no use to the farmer. It is well known, that, both in extreme hot or in extreme cold weather, the sheep always have recourse to the woods for shelter.

Account of PLANTATIONS *upon the estate of his Grace the Duke of Portland, by Mr. Speechly, gardener to his Grace. Extracted (by permission) from Dr. Hunter's edition of Evelyn's Sylva, page* 90.

FEW noblemen plant more than his Grace the Duke of Portland; and I think I may say without vanity, none with greater success. But as no man should think of planting in the very extensive manner that we do, before he is provided with well stocked nurseries, it may not be amiss before I proceed further, to give a short sketch of that necessary business; as also to inform you of the soil and situation of our seat of planting. The greatest part of our plantations is on that soil, which, in Nottinghamshire, is generally distinguished by the name of forest land. It is a continuation of hills and dales. In some places the hills are very steep and high; but in general the ascents are gentle and easy.

The soil is composed of a mixture of sand and gravel; the hills abound most with the latter, and the vallies with the former, as the smaller particles are by the wind and rain brought from time to time from the high grounds to the lower. It is on the hilly grounds that we make our plantations, which in time will make the vallies of much greater value, on account of the shelter they will afford.

After his Grace has fixed on such a part of this forest as he intends to have planted, some well situated valley is chosen (as near the centre of the intended plantations as may be) for the purpose of a nursery. If this valley is surrounded with hills on all sides but the south, so much the better.

After having allotted a piece of ground, consisting of as many acres as is convenient for the purpose, it is fenced about in such a manner as to keep out all obnoxious animals. At either end of the nursery are large boarded gates, as also a walk down the middle, wide enough to admit carriages to go through, which we find exceedingly convenient when we remove the young trees from thence to the plantations. After the fence is completed the whole is trenched, (except the walk in the middle) about twenty inches deep, which may be done for about 3l. 10s. or 4l. per acre, according as the land is more or less gravelly. This work is best done in the spring, when the planting season is over. If after the trenching, two or three chaldron of lime be laid on an acre, the land will produce an excellent crop either of cabbages or turnips, which being eaten off by sheep in the autumn, will make the land in fine order for all sorts of tree seeds: but as the oak is the sort of tree we cultivate in general, I shall confine myself particularly to our present method of raising and managing that most valuable species. In the autumn, after the cabbage or turnips are eaten off, the ground will require nothing more than common digging. So soon as the acorns fall, after being provided with a good quantity, we sow them in the following manner: draw drills with a hoe in the same manner as is practised for pease, and sow the acorns therein so thick as nearly to touch each other, and leave the space of one foot between row and row, and between every fifth row leave the space of two feet for the alleys. While the acorns are in the ground, great care must be taken to keep them from vermin, which would very often make great havock amongst the beds, if not timely prevented. Let this caution serve for most other sorts of tree seeds.

After the acorns are come up, the beds will require only to be kept clean from weeds till they want thinning; and as the plants frequently grow more in one wet season, where the soil is tolerably good, than in two dry ones, where the soil is indifferent, the time for doing this is best ascertained by observing when the tops of the rows meet.

Our rule is to thin them then, which we do by taking away one row on each side the middlemost, which leaves the remaining three rows the same distance apart as the breadth of the alleys. In taking up these rows we ought to be anxiously careful neither to injure the plants removed, nor of those left on each side. The rest of the young oaks being now left in rows at two feet apart, we let them again stand till their tops meet, then take up every other row, and leave the rest in rows four feet asunder, till they arrive to the height of about five feet; which is full as large a size as we ever wish to plant. In taking up the two last sizes, our method is to dig a trench at the end of each row, full two feet deep, then undermine the plants, and let them fall into the trench with their roots entire; and here let me observe, that much, very much, of their future success depends on this point, of their being well taken up. I declare that I should form greater hopes from one hundred plants well taken up and planted, than from ten times that number taken up and planted in a random manner; besides, the loss of the plants makes the worst method the most expensive. But before I leave this account of our method of raising oaks, I shall just beg leave to observe, that we are not very particular in the choice of acorns; in my own opinion, it matters not from what sort of tree the acorns are gathered, provided they are good; for although there seems to be a variety of the English oak, in respect to the form of the leaf and fruit, also their coming into leaf at different seasons, with some other marks of distinction; yet I am of an opinion that they will all make good timber trees, if properly managed. It is natural to suppose that a tree will grow low and spreading in a hedge row; on the contrary, it is very improbable that many should grow so in a thick wood, where in general they draw one another up strait and tall; and I have observed, that the same distinctions hold good amongst our large timber trees in the woods, as in the low spreading oaks in the hedge-row.

Though I have not as yet taken notice of any other sort of tree but the oak, yet we have a great regard for, and raise great quatities of beech, larch, Spanish chesnut, Weymouth pine, and all sorts of firs, the Scotch excepted, as well as many other kinds, by way of thickening the plantations, while young; among which the birch has hitherto been in the greatest estimation, it being a quick growing tree, and taking the lead of most other sorts on our poor forest hills; and as we have an inexhaustible spring of them in the woods, where they rise of themselves in abundance from seed, we at all times plant them plentifully, of different sizes. As to the elm and ash, we plant but few of them on the forest, though we raise great quantities of both; but particularly the ash, which being a useful wood, (but a bad neighbour amongst the oaks) we plant in places apart by itself. I shall dismiss this subject concerning the management of our nurseries, after saying a word or two relating to pruning. We go over the whole of the young trees in the nursery every winter, but in this we do little more than shorten the strong side shoots, and take off one of all such as have double leads.

Having thus pointed out the mode of forming and managing our nurseries, I shall now proceed to the plantations. The size of the plantations, at first beginning, must be in proportion to the stock of young trees in the nursery; for to undertake to plant more ground than we have young trees to go through with, for thick plantations, would turn to poor account on our forest hills. We always plant thick, as well as sow plentifully at the same time, provided it be a season in which acorns can be had, so that all our plantations answer in a few years as nurseries to succeeding plantations.

As to the form of the plantations they are very irregular. We sometimes follow a chain of hills to a very great distance, so that what we plant in one season, which perhaps is sixty, eighty, and sometimes one hundred acres, is no more than a part of one great design.

If the ground intended to be planted, has not already been got into order for that purpose, it should be fenced about at least a twelvemonth before it is wanted to plant on, and immediately got into order for a crop of turnips; two chaldrons of lime being laid on an acre will be of great service, as it will not only be a means of procuring a better crop of turnips, but will bind the land afterwards, and make it fall heavy, which is of great use when it comes to be planted, as some of the forest land is so exceedingly light, as to be liable to be blown from the roots of the young trees after planting; therefore we find it to be in best order for planting about two years after it has been ploughed up from pasture, before the turf is too far gone to a state of decay. It will be necessary to have a part of the turnips eaten off soon in the autumn, in order to get the ground into readiness for early planting, for we find the forward planting generally succeeds the best.

After the turnips are eaten off, we plough the ground with a double furrow trenching plough, made for that purpose, which, drawn by six horses, turns up the ground completely to the depth of twelve or thirteen inches. This deep ploughing is of great service to the plants at the first, and also saves a great deal of trouble in making the holes. After the ploughing is finished, we divide the ground into quarters, for the planting by ridings. It will be a difficult matter to describe the laying out the ground for this purpose, especially where there is such a variety of land as we have on the forest; much depends on the taste of the person employed in this office. Between the hills, towards the outsides of the plantations, we frequently leave the ridings from sixty to an hundred yards in breadth, and contract them towards the middle of the woods, to the breadth of ten or twelve yards; and on the tops of the hills, where there are plains, we frequently leave lawns of an acre or

two, which makes a pleasing variety. In some of them
we plant the cedar of Libanus at good distances, so as to
form irregular groves, and this sort of tree seems to thrive to
admiration on the forest-land. On the outsides of the
woods, next to the ridings, we plant ever-greens, as hol-
lies, laurels, yew, junipers, &c. and these we dispose of in
patches; sometimes the several sorts entire, at other times
we intermix them for variety, but not so as to make a regular screen or edging. Our design, in the distribution of
these plants, is to make the outsides of the woods appear as
if scalloped with ever-greens, intermixed sometimes with
rare trees, as the *liliadendron tulipifera*, or Virginian tulip
tree, &c.

After the ground is laid out into quarters for planting,
we assign certain parts to beech, larch, Spanish chesnuts,
&c. these we plant in irregular patches, here and there,
throughout the plantations, which, when the trees are in
leaf, have the most pleasing effect, on account of the
diversity of shades; especially in such parts of the forest
where, four, five, and sometimes more of the large hill
points meet in the same valley, and tend as it were to the
same center. After those patches are planted, or marked
out for that purpose, we then proceed to the planting in
general. We always begin with planting the largest
young trees of every sort, and end our work with those
of the smallest size: were we to proceed otherwise, the
making a hole for a large sized tree, after the small ones
are thick planted, would cause the greatest confusion.
Birch is generally the sort of tree we make our beginning
with, which we find will bear to be removed with great
safety, at the height of six or seven feet, though we commonly plant rather under, than at that size. This sort of
tree we are always supplied with from our plantations, of
five or six years growth. But before I proceed to the
taking them up, it will be proper to inform you, that in
the planting season, we divide our hands into four classes,

which we term makers-up, pruners, carriers, and planters; and here I shall describe the several methods of doing this work. First, in taking up we have the same care to take up with good roots in the plantations, as was recommended in the nursery, though we cannot pursue the same method; but in both places, so soon as the plants are taken up, we bed them in the ground, in the following manner: dig a trench at least fifteen inches deep, and set the young trees therein, with their tops aslant, covering their roots well as we go along, and almost half way up the stem of the plants, with the earth that comes out of a second trench, which we fill in the like manner, and so proceed on, till we have a load, more or less, in a heap, as may be convenient to the place from whence they were taken. In our light soil this trouble is but little, and we always have our plants secure, both from their roots drying, and their suffering by frost. We have a low wheeled waggon to carry them from the heaps, where they are bedded, to the pruners, and generally take two loads every other day: when they arrive, the planters, pruners, &c. all assist to bed them there, in the same manner as before described. We have a portable shed for the pruners to work under, which is also convenient for the rest of the work people to take shelter under in stormy weather. From the above heaps, the plants are taken only so fast as they are wanted for pruning, which work we thus perform: Cut off all the branches close to the stem, to about half the height of the plant, shortening the rest of the top to a conical form, in proportion to the size of the plant, and in pruning of the roots we only cut off the extreme parts that have been bruised by the taking up, or such as have been damaged by accident, wishing at all times to plant with as much root as can be had.

As soon as they are pruned, they are taken to the planters by the carriers, who are generally a set of boys, with some of the worst of the labourers: the planters go in pairs; one makes the holes, and the other sets and treads the plants fast, which work they commonly do by turns. In making of the holes, we always take care to throw out all the bad soil that comes from the bottom: if the planting be on the side of a hill, we lay the bad soil on the lower side of the hole, so as to form a kind of bason; for without this care our plants would lose the advantage of such rains as fall hastily. We at all times make the holes sufficiently large, which is done with great ease after our deep ploughing.

Before we set the plant, we throw a few spadefuls of the top soil in the hole, setting the plant thereon with its top rather inclining to the west; then fill up the hole with the best top soil, taking care that it closes well with the roots, leaving no part hollow: when the hole is well filled up, one of the planters treads and fastens the tree firmly with his feet, while his partner proceeds to make the next hole.

The fastening a tree well is a material article in planting: for if it once becomes loose, the continual motion which the wind occasions, is sure to destroy the fibres as fast as they are procured, which must end in the destruction of the plant if not prevented. It is to guard against this inconveniency, that we take off so much of the top, as has been described in the article of pruning. We plant about three or four hundred birches of the large size on an acre, and nearly the same number of the first sized oaks; we also plant here and there a beech, larch, Spanish chesnut, &c. exclusive of the patches of the said sorts of trees before planted. We then proceed to plant plentifully of the second and lesser sized oaks; and last of all, a great number of the small birches, which are procured from the woods at about three shillings or three shillings and sixpence per thousand: these we remove to the succeding plantations after five or six years. Of the several sizes of the different kinds of trees, we generally plant upwards of

2000 plants upon an acre of land, all in an irregular manner. After the planting is finished we then sow the acorns (provided it be a season that they can be had) all over the plantation, except amongst the beech, larch, &c. in the aforesaid patches. Great care should be taken to preserve the acorns intended for this purpose, as they are very subject to sprout, especially soon after gathering; the best method is to lay them thin in a dry airy place, and give them frequent turnings. We sow these acorns in short drills of about a foot in length, which work is done very readily by two men, one with the acorns, the other with a hoe, for the purpose of making the drills and covering the seed. We are of opinion, that the plants produced from these acorns, will at last make the best trees; however, I will not pretend to say how that may be, as the oaks transplanted small, grow equally well for a number of years. But it is probable that a tree with its tap root undisturbed, may in the end grow to a larger size. After the whole is finished to a convenient distance round the p.uners, we then remove their shed to a second station, and there proceed in the like manner, and so on till the whole be finished. It would be well to get the planting done by the end of February, especially for trees of the deciduous kind; but from the disappointment we met with, occasioned by the weather, we are sometimes detained to a later season. I have several times made trial of twelve or fourteen kinds of American oaks sent over to his Grace in great quantities. I sowed them in the nursery, and also in the best and most sheltered part of the plantations. In both places they come up very plentifully, but I now find that several of the sorts will not stand the severity of our winters; and those that do, make so small a progress as to promise no other encouragement than to be kept as curiosities. Towards the end of April, when the ground is moist, it will be of great service to go over the whole plantations, and

F

fasten all such trees as are become loose since their planting. After this, nothing will be more required till the month of June, when we again go over the whole with hoes, cutting off only the tall growing weeds; for the sooner the ground gets covered with grass, in our light soil, so much the better. I own there is something slovenly in the appearance of this method; and, on some lands, I would recommend keeping the ground clean hoed, for some time at first; as also planting in rows, which in that case would be necessary. More than once I have tried this method on our forest hills, and always found after every hoeing that the soil was taken away by the succeeding winds into the vallies. Besides this inconvenience, the reflection of our sandy soil is so very great, that we find the plants stand a dry season much better in our present method than in the former; and whoever fancies that grass will choak and destroy seedling oaks, will, after a few years trial, find himself agreeably mistaken: I have even recommended the sowing the poorer parts of the hills with furze or whin seed, as soon as they are planted. We have sometimes permitted the furze to grow in the plantations, by way of shelter for the game, which, though it seems to choak and overgrow the oaks for some time, yet after a few years, we commonly find the best plants in the strongest beds of whins. This shews how acceptable shelter is to the oak whilst young; and experience shews us, that the oak would make but a slow progress on the forest hills for a number of years at the first, were it not for some kind nurses; and the birch seems to answer that purpose the best, as I have already observed. The several sorts of fir trees, from appearance, seem to promise a great shelter; but on the forest land they do not grow so fast as the former; and what is worse, the oak will not thrive under them, as they do immediately under the birch. Where a plantation is on a plain, a screen of firs for its boundaries is of singular use, but the situation of the forest land denies us this advantage. We continue to cut down the tall growing weeds two or three times the first summer, and perhaps once the next, or second season after planting; which is all that we do in respect to cleaning. The next winter after planting, we fill up the places with fresh plants where they have miscarried, after which there is little to be done till about the fourth or fifth year, by which time the small sized birch, and seedling oaks, will be grown to a proper size for transplanting. In the thinning of these, due care must be had not to take too many away in one season; but being properly managed, there will be a supply of plants for at least half a dozen years to come. About the same time that the lesser sized birch wants thinning, the large ones will require to have their lower branches taken off, so as to keep them from injuring the oaks, and this is the first profit of our plantations; the birch wood being readily bought up by the broom-makers. This pruning we continue as often as required, till the birches are grown to a sufficient size to make rails for fencing; we then cut them down to make room for their betters. By this time the oaks will be grown to the height of twelve or fourteen feet, when they draw themselves up exceedingly fast. Each plant seems as it were in a state of strife with its neighbour, and in a strict sense they are so, and on no other terms than life for life: and he whose fate it is to be once over-topped, is soon after compelled to give up the contest for ever.

After the birches are cut down, there is nothing more to do but thinning the oaks, from time to time, as may be required, and cutting off their dead branches as frequently as may be necessary. We are very cautious in doing the former, knowing well that if we can but once obtain length of timber, time itself will bring it into thickness; therefore we let them grow very close together for the first fifty years. And here it may not be improper to observe the progress the oak makes with us, by describing them in two of our plantations, one of twenty-eight, the other of fifty years growth. In the former they were in general about twenty-five or twenty-six feet in height, and in girth about eighteen inches. The trees in the latter, planted in 1725, are something more than sixty feet in height, and in girth a little above three feet, and these trees are in general about fifty feet in the bole, from which you will easily conceive the smallness of their tops even at this age. It would be a difficult matter to describe their farther progress with any degree of certainty; therefore let it suffice to make this last observation on them in their mature state. I should have observed to you, that in both the aforesaid, as well as in all the young plantations, the Spanish chesnut keeps an equal pace, or rather outgrows the oak; but it is doubtful whether ever they will arrive at the same size; for the largest of our Spanish chesnuts, which have much the appearance of old trees, do not girth more than twelve or fourteen feet, which is nothing in comparison to some of our large oaks, which girth from twenty-five to thirty feet; indeed some of them a great deal more: for instance, that remarkable tree called the Green Dale Oak, (from its growing in a valley of that name near Welbeck) which in the year 1724 had a hole cut through its body large enough to admit a coach to go through. This great curosity is yet living, and frequently bears acorns, which we carefully save, to be distributed as presents amongst his Grace's acquaintance. I may not omit describing to you the present state of this piece of antiquity, as I have herewith inclosed a drawing of it taken on the spot a few days ago, from which you will see, notwithstanding the uncommon size of the lower part of the tree, that it has never contained any great quantity of timber; I mean in comparison with several of our largest oaks; some of which contain, in their tower-like trunks, between seven and eight hundred solid feet of timber, exclusive of their stately tops; and some of their large branches are even like trees themselves.

You see, Sir, what a surprising mass of wood may arise from a single acorn: indeed it is really wonderful to see, on some soils, to what an amazing size this king of trees will sometimes arrive.

Welbeck, June 16, 1775.

The uncultivated lands which have been improved by agriculture and planting, by the Duke of Portland, within the last twenty-five years, amount to between two thousand and two thousand five hundred acres; of which number about one fourth consists in plantations.

Woods and **Plantations** *in the district of the forest and borders, the greater part within thirty years.*

		Acres
Martin Woods	Duke of Newcastle's (oak spring, few poles good) timber getting up	190
Mattersey Woods	Jon. Acklom, Esq. and others	50
Worksop Manor	Duke of Norfolk's	781
Welbeck Park	Duke of Portland's near	400
Carberton, &c.	Ditto - - about	600
Clumber Park	Duke of Newcastle's	1848
Apley Head, Bothamsell, &c.	In this district, ditto	349
Blyth, &c.	Charles Mellish, Esq. chiefly fir between 300 and	400
Serlby and Farworth	Lord V. Galway	136
Scofton	Robert Sutton, Esq. about 60 fir, same of old standing, very large, the rest oak and beech about	200
Osberton	F. F. Foljambe, Esq.	184
Rufford, &c.	Hon. R. Lumley Savile (in this district)	491
Thoresby, &c.	Chs. Pierrepont, Esq.	981
Wiseton	Jonathan Acklom, Esq.	25
Carlton, in Lindrick	Robert Ramsden, Esq. upon strong soil, and bog earth, oak, Spanish chesnut, and ash	148
	Forest plantations	48
Ditto	Charles Mellish, Esq.	154
Ditto	Taylor and Wollaston White, Esq.	107
Ditto	Mr. Joseph Cowlishaw	15
Wallingwells	Taylor White, Esq.	28
Linby and Papplewick	Right Hon. Fred. Montagu oak, ash, elm, &c. 40 acres in preparation	34
Farnsfield	Sir Rich. Sutton's	20
Ditto	Robert Lowe, Esq.	18

There are besides these a number of dispersed clumps, and plantations of smaller extent.

An Account of His Grace the Duke of NEWCASTLE'*s* ASH PLANTATIONS, *and a Valuation of the Lands before they were planted.*

COMMUNICATED BY WILLIAM MASON, ESQ.

	A.	R.	P.	Value per Acre. s. d.	Yearly Value. £. s. d.
HAUGHTON.					
Bog Close, late Geo. Padley	10	0	33	8 0	4 1 2
Ditto, late Dewick	20	0	14	2 0	2 0 5
Great Kennel in Dewick's bogs	3	2	0	7 0	1 4 6
Little Kennel, in ditto	2	0	0	5 0	0 10 0
Dobson Hopyard	1	0	0	10 0	0 10 0
Gosling Carr	8	1	37	5 0	2 2 0
Crow Park	5	2	0	9 0	2 9 6
Decoy Plantation, Including the banks about Fat Close Ponds	7	2	0	15 0	5 12 6
Cross Close Plantation	3	2	0	12 0	2 2 0
Bog ground	61	3	13		20 12 0
BEVERCOTES.					
Farny's Plantation	51	1	0	8 0	20 10 0
Clay ground					
WALESBY.					
Some pieces of land laid to Bevercotes wood	1	0	27	13 0	0 14 0
A piece of land laid to Nichhaghbushwood	1	2	32	10 0	0 17 0
Clay ground	2	3	19		1 11 0

	A.	R.	P.	Value per Acre. s. d.	Yearly Value. £. s. d.
West Drayton.					
Schol-house Close Plantation	3	0	0	8 0	1 4 0
Bog ground					
Gamston.					
Land adjoining the wood	2	2	0	5 0	0 12 6
Ditto upon the common	5	0	0	10 0	2 10 0
Clay ground	7	2	0		3 2 6
In Bothamsell and Elksley.					
In Patmoor	41	0	0	5 0	10 5 0
Crookford	14	0	0	5 0	3 10 0
In Eiksley hop-yards by the fish pond below Elksley wood	0	2	0	10 0	0 5 0
Bog ground	55	2	0		14 0 0
Ash Plantations collected.					
In Haughton	61	3	13	0 0	20 12 0
Bevercotes	51	1	0	0 0	20 10 0
Walesby	2	3	19	0 0	1 11 0
West Drayton	3	0	0	0 0	1 4 0
Gamston	7	2	0	0 0	3 2 6
Bothamsell and Elksley	55	2	0	0 0	14 0 0
	181	3	32		60 19 6
The above 180 acres, now planted, are worth upon a moderate calculation, 40s. an acre	180	0	0	40 0	360 0 0

The expence of planting the above 180 acres, at 15l. an acre, amounts to 2,700l. for which his Grace will have an additional rent of 300l. a year.

OF NOTTINGHAMSHIRE.

PLANTATIONS OF OAK.

	A.	R.	P.
Mr. Ormand's hill at Gamston	2	2	0
In Haughton park, by Elksley wood	10	0	0
In Walesby warren, by Gosling Carr	1	1	0
	13	3	0

OTHER PLANTATIONS.

In HAUGHTON.

	A.	R.	P.
In Richard Rawson's warren	1	0	0
At Haughton chapel	1	0	0
In Park Meadows	2	2	0

IN BOTHAMSELL.

	A.	R.	P.			
In Peck's farm, two plantations	1	0	0			
In Dewick's farm, one ditto	3	2	0			
In Moss's farm, Broom plantation	1	2	36			
Border, between Moss and Mr. Bower	5	1	32			
Fivethorns plantation	0	3	0			
In Wood's farm	2	2	0			
In Hunt's	0	2	0			
In Mrs. Padley's	4	0	0			
Hill above Ousedale, from Padley's watering place, to Normanton coach road gate	16	3	5			
				38	0	33

IN NORMANTON.

	A.	R.	P.			
Border, from the coach road gate, by Mr. Bower's farm, to the house	16	2	2			
From the stack-yard, along the west road, down to the white gate	7	2	8			
Below the above white gate, next the bottoms	11	3	34			
Large round clump in the low westernmost piece	7	0	24			
Three small rounds in the piece east of the above	3	0	35			
South and east side of the above piece	7	2	20			
Kennel plantation in the rye piece	3	1	0			
Small border south of the lane	1	1	16			
Some other pieces of planting, suppose	10	0	0			
				68	2	19

	A.	R.	P.
Brought over	63	3	19

OTHER PIECES TAKEN IN TO BE PLANTED.

	A.	R.	P.
One piece	8	2	32
Another	15	0	8
Another	48	2	10
	72	1	10
Deduct as above, it being included in the above three pieces	10	0	0
	62	1	10
Aphyhead	139	0	36
Deduct for the six avenues, suppose six acres each	36	0	0
				103	0	36
				214	2	8

N. B. All these plantations, amounting to 410 acres 1 rood, have been made within twenty years.

The account of the WOODS *and* PLANTATIONS *of the Hon.* R. LUMLEY SAVILE, *Esq. is contained in the following letter:*

Rufford, Jan. 28, 1794,

SIR RICHARD,

ACCORDING to my promise, I send you the number of acres of the Hon. R. Lumley Savile's woods and plantations, which is the best account he has; but don't pretend to say it is exact, for want of surveys of part of them. I have also sent you some observations about the woods, and some other things.

Part of these are oak woods; the rest are oaks, and planted with ashes for hop-poles.

	Acres.	
Wellow Park, and Birkhill	266	clay
Egmanton wood	100	clay
Eakring Brail	52	clay
Lound wood in Rufford	24	clay
The rest of the woods in the liberty of Rufford	200	100 cl.
Bilsthorp woods	22	clay
Worney wood	35	clay
Oak plantations in Rufford	50	sand
Pittance Park plantation in ditto part oak and the rest firs	218	ditto
Rufford fir plantations	50	ditto
Bilsthorp ditto	50	ditto
Ollerton hills ditto	23	ditto
	1090	

Woods require great attention to make them profitable. In the first mentioned woods, which are oak timber, and young oaks, they have been thinned, and most other trees cut down to make room for the oaks, where they were likely to make good trees. They have been thinned by degrees, which draws them up high enough, and gives them room to thicken. There are a few Spanish (or the best) chesnut trees, which is a quick growing wood, and grows tall and strait; from experiments made about twenty-five years ago, it is found to be excellent wood, much resembling oak, both in colour and quantity.

In those woods, which are oaks, &c. and hop-poles, they are cut about every twenty years, leaving the small oaks, and here and there an ash, elm, or beech, &c. The largest oaks, worth about three pound and upwards, are taken down; also all the crooked bad ones. The hop-poles are cut off neatly, and the year following the vacant places filled up with ashes; the crooked shoots are pruned off, and any small bad oaks, damaged with falling, cut down. Great care is taken to keep cattle out of these woods.

In the oak plantations the land was fallowed, and the greatest part of it sown with acorns about forty-four years ago. Some of them were sown in drills, and ploughed betwixt, which answers the best. The trees were pruned and thinned about twenty years ago, and are now nearly ready for pruning again. These oaks are now about twenty feet long, are from twelve to twenty-four inches in girt, and are thriving oaks. Some oaks were planted at that time, but there is little difference betwixt those planted, and that sown with acorns; if there be any difference, those raised from acorns are the best. Two extremes ought to be avoided; not to thin them too much, till they have grown a sufficient height, nor leave them too thick to kill one another.

A considerable part, Sir, of those plantations were planted with firs, by the late Sir George Savile, Bart. at the same time the acorns were sown, viz. about forty-four years ago. These firs have made great improvement. Larch is the quickest growing wood of all firs, and is the most useful. Scotch and silver fir seem to be the next. Spruce grows more slowly than any of them. The larch now measure from thirty to fifty feet long, and from twelve, to forty-eight inches in girth. The Scotch are from twenty to thirty feet long, and are from twelve to thirty-two inches in girth. They were planted rather shallow, were pruned and thinned for hop-poles, when about twenty-five years old; at thirty were fit for fencing, and now at forty years old, they are fit for building timber, fencing stuff, and some of them boards. From experiments made of some older firs, it is found that larch is the best of them all for enduring, especially in wet. The others will last very long if kept dry. Some of them were planted in tribes, which seem to answer the best. Firs ought always to be planted by themselves, as they are apt to kill most other trees near them. All firs ought to be felled in summer, because they are then fullest of turpentine, and heaviest. If furs are suffered to grow too thick till they are a great height, they will die in patches, after they have exhausted the ground; and if they are thinned too soon, they will never rise a great height. Upon the whole, firs are likely to prove of more value to this country than was expected some years ago.

The late Sir George Savile inclosed the following lands, since the year 1776:

	ACRES
Ollerton forest, about	500
Bilsthorp clay fields	100
Ditto commons	280
Morton Grange forest	1080

There are yet open fields at Eakring, Ompton, Wellow, Laxton, Egmanton, Walesby, and Cirton.

Rufford is now almost the only uncultivated forest Mr. Savile has in this county.

Improvements, Sir, in husbandry are best promoted by example; for one capital manager living in a neighbourhood is, or might be, a benefit to all near him.

The drill husbandry for beans and pease is particularly useful, as it prepares the land for wheat.

If the same attention was paid to the breeding of horses, beasts and pigs, that is paid to the breeding of sheep, it would be of great service to this country.

Growing carrots for horses, and potatoes for milch cows (which produce sweet butter) are among the newest Improvements here. Potatoes are of the greatest use, cheap and wholesome: the poor have never suffered so much from want since they have been so generally cultivated.

  I am, Sir Richard,
    with the greatest respect,
      your most obedient humble servant,
          JOHN PARKINSON.

Sir Richard Sutton, Bart.

Account of the WOODS *and* PLANTATIONS *of* CHARLES PIERREPONT, Esq. *now* Lord NEWARK.

Whitemoor, January 21, 1794.

SIR,

  HEREWITH you will receive a copy of Mr. Calvert's measure of the different plantations in and about Thoresby Park, returned to me yesterday. I have collected into columns the best and the worst sort of trees, and the number of acres of each, which I should think would be sufficient, rather than to insert the names of such a numerous quantity of clumps.

  I forgot to mention in my paper respecting Mr. Pierrepont's spring woods, that we have adopted a plan of filling up every fall, the following spring after it is cut, with ash plants, all the vacant places; and I think it a good method, where a kind young ash tree was left the fall before (for a husbandry pole) and beart keys, to leave it the second fall, and if you see occasion the third, that the wind might disperse the keys, and be a means of replenishing the woods with ash plants without any expence.

    I am, Sir,
      your obedient servant,
         W. PICKIN.

To Sir R. Sutton, Bart.

PLANTATIONS *in and about* THORESBY PARK, *belonging to* CHARLES PIERREPONT, *Esq.*

	A.	R.	P.
Oak, ash, beech, Spanish chesnut, and elms	797	1	13
Firs of different kinds, and birch	148	0	13
Ashes, ollers, willows, and sallows	36	0	27
Total	981	0	0

N. B. There are 306 acres now in preparation for planting.

Mr. PIERREPONT's WOODS *in the County of* Nottingham. *Communicated by Mr.* PICKIN, *Steward to Mr.* PIERREPONT.

[The letters denote the districts; L Limestone, CL Clay, N. of Trent.]

	A.	R.	P.
Kirkby (L) Coalfield-wood	8	0	0
Holbeck (L) Old Hag wood	20	0	0
Lexington, (CL) East Park-wood	31	0	0
Ditto Saw-wood	38	0	0
Ditto Middle Spring	45	0	0
Knesal (CL) Green-wood	64	0	0
Eakring (CL) Hobheron-wood	19	0	0
Brail-wood	29	0	0
White-stubbs	12	0	0
Weston (CL) Lady-wood	15	0	0
Gedling (CL) Marshall hills, about	100	0	0
ACRES	381	0	0

N. B. The above woods (excepting the last) are spring woods, and cut about every fifteen years, at about eighteen acres per annum; the underwood consists of hop-poles, husbandry-poles, and hedge wood; and the weeding of the oaks generally applied in the reparations of the buildings upon the estate.

Water Meadows.—About twelve acres in Whitemoor Farm can be well watered, and with very good effect.

Mr. Pierrepont has about twenty acres of boggy ground capable of that improvement.

An Account of PLANTATIONS *at* Osberton, *belonging to* F. F. FOLJAMBE, Esq.

Rufwood	-	7 acres, new planted.
Crowood	-	6 ditto
Broomwood	-	14 ditto
Whincover	-	12 ditto, new planted.
Grote planting	-	14 ditto
Grote border	-	30 ditto, new planted.
Square wood	-	8 ditto
Side border	-	16 ditto, new planted.
Ash wood	-	12 ditto
Twenty acres wood		8 ditto
Danbottom wood		5 ditto
Yewtree wood	-	5 ditto, new planted.
Spring wood	-	10 ditto
Parks, ditto	-	7 ditto
Keniwell spring		30 ditto, new planted.
		184

An Account of Lands planted in the Parish of Carlton, *in* Lindrick, *in the County of* Nottingham, *as follows:*

CHARLES MELLISH, ESQ.

	A.	R.	P.
Great plantation	100	0	0
Five other plantations	54	2	0
	154	2	0

TAYLOR WHITE AND WOLLASTON WHITE, ESQS.

	A.	R.	P.
Plantations in the Calf Spring	2	2	0
Carlton wood	52	2	0
Coal Pit wood	8	2	0
Plantations below Holme field	0	2	0
Owday wood	43	3	0
	107	3	0

COWLISHAW JOSEPH.

	A.	R.	P.
Plantation	15	1	0

Letter to WILLIAM MASON, *Esq. of* WELHAM.

SIR,

I FIND from the person who overlooks Mr. White's woods, that Mr. Edwards (when last down) took with him all the books and plans respecting the estate at Wallingwells; but that you might not be disappointed in your inquiries, I the other evening took a survey of the undermentioned woods, which is much at your service.

I am, Sir,
  your most obedient and
    very humble servant,
      JOS. YOUNG.

Carlton, Feb.

WOODS *and* PLANTATIONS *belonging to* TAYLOR WHITE, *Esq. at* WALLINGWELLS, *in the County of* NOTTINGHAM.

	A.	R.	P.
Lingby wood	22	0	0
Plantation north of the park	2	3	0
One small plantation near the last	0	3	0
New planted in the park, about	3	0	0
	28	2	0

A Particular of WOODS *and* PLANTATIONS *belonging to His Grace the Duke of* NORFOLK, *at and near* Worksop Manor, *in* Com. Notts, *viz.*

Munton Plantations,

	A.	R.	P.
Near Worksop, planted from 1762 to 1764, oak, ash, larch, Scotch, and other firs, (sand)	168	0	0

Sparking-Hill Wood,

	A.	R.	P.
Near Worksop, (sandy soil) planted in 1762, oak and chesnut	15	0	0

Hannah Park,

	A.	R.	P.
Near Worksop (sandy soil) planted in 1744, beech and firs	8	0	0

Kilton Plantations,

	A.	R.	P.
Near Worksop (sandy soil) planted in 1763, oaks chiefly	36	0	0

Tranker Wood,

	A.	R.	P.
Near Worksop (soil clay) planted about the year 1727, oak with beech intermixed	10	0	0

Tranker Wood, in dispute with MR. HOWELL,

	A.	R.	P.
Near Worksop (clay soil) principally oak of very ancient growth	60	0	0

Clumps,

	A.	R.	P.
In the park and on the waste (sandy soil) oak, larch, firs of all sorts; elm, beech, and chesnut, planted in 1758, and subsequent years	34	0	0

Small Plantations of Spring Wood,

	A.	R.	P.
Near Worksop, different soils	16	0	0

The Menagery,

	A.	R.	P.
A plantation of oak, elm, chesnut, beech, cherry, larch, Scotch, and other firs; planted in 1733, and subsequent years, (sandy soil)	20	0	0

The Wilderness,

Near the Manor House, planted at different times from 1713 to 1738; soil clay in part, and part sand; consists of oak mostly, but intermixed with elm, beech, chesnut, larches, pines, Scotch and other firs of great magnitude - - - - 58 0 0

Manor-Hill Plantations,

Were in great part planted in 1734, and subsequent years (soil sandy) consist of various sorts of wood, viz. oak in great quantity; beech, elms, chesnut, pines, larches, spruce, Scotch, and other firs, so disposed as to form a pleasing mixture of variety to the distant view; contains in the whole - 350 0 0

An Oak Wood,

At Stirrup, of ancient growth - - 6 0 0

Total of woods in Com. Notts   781 0 0

There is little wood *in the Trent Bank District*, but in hedge rows.—In 1771 and 1772, the cliff opposite Haxleford Ferry, in Flintham, which had formerly been covered with good timber, was planted was ashes for springing. The extent thirty-one acres, three roods, fourteen poles. They were cut in 1791 and 1792. Thirteen hundred pounds were offered for them standing in 1791, making above forty shillings an acre for the time of growth. Five acres, one rood, eight perches, were planted in 1790. In Kneeton, in 1781, six acres of ash wood sprung seventeen years before, were sold standing for seventy pounds.

I have since obtained this further list of
WOODS in *Trent Bank District*:

LORD MIDDLETON'S AT WOLLATON.

	A.	R.	P.	
Paddock Wood	57	2	11	timber
Groups in the park	25	0	0	ditto
Shepherd's Wood	26	1	5	ditto
Little Leek, one 12, one 4	16	0	0	ditto

PLANTATIONS.

	A.	R.	P.
Little Leek	26	0	0
Wollaton park	10	0	0
Brickhill plantation	11	2	25
Dovecourt	22	0	0
Goss plantations, &c.	28	1	0

LANGFORD PLANTATION.

	A.	R.	P.
Mr. Duncomb's	10	0	0
Ditto on the moor, mostly oak	4	0	0

WINTHORP, ROGER POCKLINGTON, ESQ.

	A.	R.	P.
Ash, elm, fir, pines, larches, beeches and some oaks	38	1	25

CARLTON ON TRENT.

	A.	R.	P.
J. Pocklington, Esq. oaks	12	0	0
Firs, ash, elm	6	0	0
N. Musklam to ditto	6	0	0
Weymouth pine, spruce, larch.			

For the plantations at Thorney, in the tongue of land east of Trent, I must refer to the letter of G. Nevile, Esq. inserted before.

In the Clay District, north of Trent, there are considerable tracts of woods, which are chiefly sprung. Of these I have inserted in the following pages a general and particular account, with the management of them as far as I have been yet able to learn. I may observe in general, that the principal value of spring woods in this country, arises from the ash for hop-poles; and the stakes and bindings, flakes, &c. for farmers' use. From the universal use of coal for fuel, brush and cord wood are of less value than in many other counties; the bakers even having learnt to heat their ovens with coal. The timber in most of the woods in this district having been cut within these twenty years, and the underwood hurt by the growth of the timber, they have been reduced in value; but are now, in general, improving by new planting and taking care of the underwood, and young oaks are getting up for timber being left for standards. Vacancies are generally filled up with ash. Some charcoal is made.

Thorney Wood Chace.—A branch of the forest of Sherwood, of which the Earl of Chesterfield is hereditary keeper, by grant of 42 Eliz. comprehends most of the towns mentioned in the southern part in the survey of 1609. It is well stocked with fallow deer, as the rest of the forest was formerly with red deer, which appear not to have intermixed. It has been hitherto well wooded; but the recent inclosures of Lambley and Gedling, when completed, will reduce it to very little. *In point of soil,* only the towns of Carlton, Gedling, Burton, and Bulcote, Lowdham, Lambley, Woodborough, and part of Arnold and Calverton, fall within this district. The deer since the late inclosures are all destroyed.

General List of WOODS *and* PLANTATIONS *in the* CLAY DISTRICT NORTH OF TRENT.

	belonging to	Acres
Grove and Headon Woods	Ant. H. Eyre, Esq. about all spring woods, some at twelve or thirteen, some at eighteen years. Pay about 30l. an acre when cut. A good deal of ash and oak. Some standards for timber.	110
Ditto	Ash Plantations — cut once in fourteen or fifteen years, paying 30l. an acre.	30
Eaton	Mr. Simpson's — not many poles.	50
Treswell	Earl of Bute's — some timber, about 1000 poles an acre, 10l. an acre.	128
Gamston	Duke of Newcastle's about sprung for hop poles, not 300 per acre.	100
Bevercotes	Duke of Newcastle's	90
Nickhagh Bush	Ditto	12
Askham Wood	Robert Sutton, Esq.	12
Wheatley, &c. &c.	Lord Middleton's	191 odd
Ossington Woods	J. Denison, Esq. about	400
Wellow Park, &c. &c.	Hon. R. Lumley Savile	564
Lexington or Laxton, &c.	Charles Pierrepont, Esq.	283
Norwell	Prebend of Southwell sprung	50
		2030

	belonging to	Acres
	Brought forward	2030
Epperston	Earl Howe's	349 odd
Ditto	Other Proprietors	26
Lowdham Springs	F. W. Edge, Esq. Mr. Broughton, Mr. Wright, and Miss Briggs	126
Burton	Earl of Chesterfield timber oak, sprung once in eighteen years, chiefly hasel and thorn, timber oak and ash much decreased.	100
Eastshaw Wood	Earl of Chesterfield's	47
Winkburn Woods	W. Pegge Burnell, Esq. sprung once in twenty years, pays at least fifteen shillings per acre yearly. Formerly much neglected, now the vacancies are filling up with ash.	400
Kirklington Woods	Mrs. Whetham about small timber for fleaks, &c. Sprung about every fourteen or fifteen years. Fine poles, not many oaks; worth about sixteen shillings an acre, annually.	100
Averham Park	G. Sutton, Esq. (including thirty new planted) Sprung in fifteen or sixteen years; in the whole about sixteen shillings per acre. Good young oak and ash.	100
Comb Woods	Archbishop of York's oak trees, few of them timber, ash for poles, and brushwood.	48

3326

	belonging to	Acres
	Brought forward	3326
Halloughton	Sir R. Sutton, under Southwell prebend of Halloughton, formerly much neglected - Small oaks from stools; now bringing into regular springs; vacancies filling with ash	43
Oxton	Mrs. Sherbrooke's - ill planted	52
	Mr. Lowe's new *plantation* ten or twelve acres of Huntingdonshire willows, the rest oak, ash, &c. The willows answer by far the best.	42
Marshall Hills	Mr. Pierrepont's - open wood in Thorney Wood Chace now allotted.	100
Bevercotes, &c.	Duke of Newcastle's *plantations on clay*, -	61
		3624

Ossington, Jan. 26, 1794.

SIR,

I YESTERDAY received the letter you did me the honour to write, requesting I would furnish you with an account of the extent and manner of occupation of my woods, to assist Mr. Lowe in his survey, which I take the earliest opportunity of stating.

They are nearly four hundred acres ash and oak; about twenty acres are annually cut down for hop-poles, round poles, and country uses. Such oaks as are stag headed, and not likely to improve against another fall, are felled at the same time. The hazels and thorns are afterwards mostly stubbed up, and young ashes planted in their stead; by which mode, with the addition of draining the wet parts, these woods have been very considerably improved. I last year completed the whole round on this plan, which has taken an immense number of plants. I have some years planted from eighty to one hundred thousand, and mean still to pursue the same plan, though in a diminished proportion. Some of the young plants make poles the first fall; but in general, they are not supposed to be productive till the second, and many die. Oaks in this soil do not grow to any girth, but are mostly straight, solid, good hearted timber.

Should you or Mr. Lowe wish for any further particulars, I shall be happy to give you every further information in my power.

I am, Sir,

your most obedient humble servant,

JOHN DENISON.

To Sir Richard Sutton, Bart.

COMMUNICATED BY MR. DUFTY.

	A.	R.	P.	
Epperstone Park,	208	2	0	The whole of the under-
Lingar Wong,	5	0	13	wood is cut down once
Brockwood Hill, and				in seventeen years.
Eastwood	136	0	0	

	A.	R.	P.	A. R. P.
Lord Howe's	349	2	13	20 3 10$\frac{7}{17}$ upon average per year.

	A.	R.	
Mr. Bingham's Stubbings,	4	0	Woods belonging to other
Mr. Barnard's Wood,	2	2	proprietors in the parish
Mr. Smith's, about	10	0	of Epperstone.

N.B. These woods of Lord Howe's are begun to be filled up with ash

---

WOODS in *Thorney Wood Chace*, allotted.

In GEDLING.

		A.	R.	P.
	Podder Coppice,	55	3	20
	Leeson, ditto,	42	0	15
Allotted	Ouscah, ditto,	10	1	30
in farms,	Park Well, ditto	61	0	15
and cul-	Stone Pitt, ditto	65	3	10
tivated.	Harbor Hill, ditto	34	2	5
	Pismire Hill, ditto,	65	3	20
	Old ditto,	51	0	0

	A.	R.	P.
	396	2	35
Plains	53	2	5

	A.	
East Haw,—Earl Chesterfield,	47	Allotted, but re-
Marshall Hill,—C. Pierrepont, Esq.	100	main in wood.
	147	

In LAMBLEY.

	A.	R.	P.
Allotted and cultivated, Coppices,	408	0	0

In ARNOLD.

	A.	R.	P.
Now in tillage, Coppices,	50	0	0
Acres	854	2	35

On the Vale of Belvoir are no woods of any account.

There is very little wood in the district of the Nottinghamshire Woulds.

Bunny Woods, belonging to Sir Thomas Parkyns, Bart. are about seventy acres.

In the *Lime and Coal District* are considerable woods, as will appear by the following accounts:

Account of the WOODS, &c. *in the Lime and Coal District on the Estate of Lord* MIDDLETON, *communicated by* Mr. REYNOLDS, *Gardener to his Lordship.*

	A.	R.	P.	
Broomhill Woods	13	0	25	
Addess Wood	21	0	31	
Asply High and Low Woods	11	1	4	
At TROWEL				
Short Wood	67	0	37	All Timber.
Lawns Wood	35	1	10	
Grange Wood	12	0	30	
At BROXTON				
Slang Wood	56	1	6	
Cinderhill Wood Sprung	17	3	37	
Odworth One Small Wood	4	0	0	
Plantations				
At TROWELL				
Shaw's Hill Plantation	4	0	2	planted June 1782
Grange Wood New Plantation	4	2	1	

Further List of WOODS *in the Limestone and Coal District.*

		A.	R.	P.
Cuckney, &c.	Earl Bathurst's, vide page 2,	196	2	22
Warsop, &c.	J. G. Knight, Esq. vide p. 3,	13	2	0
William's Wood,	Chapter of Southwell	41	0	0
Greasley Woods,	Lord Viscount Melbourne	285	0	0
Selston Wood,	Ditto	12	0	0

These two last are timber woods of good oaks, some stubbed, none fresh planted: no ash.

Nuthal,	Honorable Henry Jedley	111	1	25
Watnal,	—— Roulston, Esq.	83	2	28
	Acres	916	0	24

N.B. Lord Bathurst and Mr. Knight's woods, and that of the Chapter of Southwell, standing just in the division of the Sand and Limestone, and most of them partaking of both, I have thought it better to put them here all together, than split them into the several districts.

Particulars of Mr. ROULSTON's *Woods.*

	A.	R.	P.
Thorp Wood	52	3	10
Hill Hole	14	3	33
Starth Wood	4	0	0
Leyhill Wood	16	3	25
	88	2	28

An Account of the Wood Lands belonging to the Right Honorable Earl BATHURST, in the County of Nottingham.

Wood's Names.	Quantity.			Quality.
	A.	R.	P.	CUCKNEY TOWNSHIP.
Elksley Hill	7	2	6	Planted with firs, beech, ash and birch.
Forest Head	1	2	37	Planted with firs and birch.
Church Plantation	1	0	36	Planted with firs, oaks, beech and alders.
Mill Hill	3	0	6	Planted with firs, oaks, beech and birch.
Deadman's Grave	13	0	13	Spring wood and fine oak timber.
				LANGWITH TOWNSHIP.
Cuckney Hay	145	0	17	Spring wood and fine oak timber.
Langwith Plantations	9	1	33	Planted with oaks, firs, beech, ash, and sundry ornamental trees and shrubs
Boon Hills	8	0	30	Oak, elm, ash, and lime timber, and planted with firs, ash, &c.
Lady's Grave	7	0	13	Spring wood and fine oak timber.

Letter to Sir R. SUTTON, *Bart.*

HON. SIR, *Warsop, Feb.* 4. 1794.

THE following are the accounts of Mr. KNIGHT's woods at Warsop in this county:

Names.	Extent.	
Collyer Spring,	75 ACR.	A spring wood, with fine growing oaks, underwood, as hop poles, &c.
Lord's Stubbing	37 ditto	A spring wood as above, but the oaks not so thriving.
Coppice	16 ditto	A spring wood, much as last.
Parson's Spring	13¼ ditto	A spring wood, with growing oaks and underwood as above.
Rough Wood	7¼ ditto	A spring wood, as above, but the oaks not so thriving.
Moscarr Rough Wood	2¼ ditto	A spring wood, much the same as last.

KIRKTON, *in this County:*

Norton Wood	8 ACR.	A spring wood, with fine growing oaks, underwood as hop poles, &c.
Thorney Wood	8 ditto	A spring and timber wood, with very fine growing oaks, &c. What I mean by timber, some part of this wood is rather strong growing oaks, and thick upon the ground.

WILLOWBY, *near Kirton, in this county:*

High Spring	4 ACRES.	A spring wood, with fine growing oaks, underwood, &c.

Yesterday I delivered your inclosed letter at Mr. ROBINSON's house myself, but he was not home. If, Sir, you want any further account or explanation, I shall be glad to render you every service in my power. And

I am, Sir,

Your very humble servant,

SAMUEL JACKSON.

To Sir R. SUTTON, *Bart.*

Warsop, Feb. 15, 1794.

HON. SIR,

AGREEABLE to your request, the following is the statement of the soil in the sundry woods:

Collyer Spring	Half sand, the other half loamy.
Lord's Stubbing	Woodland clay soil, bordering upon limestone.
Coppice,	Ditto
Parson's Spring,	Half limestone, the other light clay.
Rough Wood,	Limestone.
Moscarr Rough Wood,	Woodland clay.
Norton Wood, Thorney Wood, High Spring.	A good clay soil

Lord Bathurst's wood, half sand, a fourth limestone, the other fourth a loamy soil.

S. JACKSON.

CHAPTER XI.

Wastes.

LITTLE waste land is now left in this county, much the greater part of the forest being inclosed. What remains is chiefly in the line between Rufford and Mansfield, and between Blidworth and Newsted, and is mostly poor barren land.—A good deal of it is in rabbit warren; and it is to be doubted whether it would answer so well in any other shape. Some part has formerly been taken into cultivation, and thrown up again. Planting might perhaps be found to answer.

In the tongue of land east of Trent are some low flat commons, very much drowned in winter, and not easily imimprovable without good drainage, and whether the outfall is sufficient seems questionable.

The ancient royal forest of Sherwood is in extent, from Nottingham to near Worksop, about twenty-five miles; and in breadth, seven, eight, or nine miles, more or less in different places. Several tracts of land, particularly in the north part as far as Rossington bridge, lying in the same waste state, have been usually called forest; but from the survey of 1609, appear not to have belonged to the forest, or to have been disafforested before that time. In it are comprehended several parks taken in at different times, as Welbeck, Clumber, Thoresby, Beskwood, Newsted, Clipston, and several villages or lands belonging to them.* The whole soil of the forest is understood to have been granted off from the Crown to different lords of manors, reserving only, in forest language, the vert and venison, or trees and deer.

The latter were formerly very numerous, all of the red kind. Within the memory of persons living, herds of one hundred or more might be seen together; but as cultivation increased, they diminished gradually, and are now entirely extirpated. The vert and venison are under the care of four verdurers, chosen by the freeholders of the county.

* Vide Villages in the forest and extent and offices of the forest, Appendix, No. VIII.

CHAPTER XII.

Improvements.

SECTION I.

DRAINING.

Drainage.—THE necessity of draining wet lands, has of late years been much better understood and attended to, than formerly, and the rot amongst sheep of 1792 has alarmed, and almost every where brought forth exertion in this respect.

In the new inclosure bills, drains are ordered by the commissioners, and provision made for their being properly kept up, which is more effectual than the old laws of sewers, of the neglect in the execution of which there is great complaint here, as well as in other counties.

The drainage has been facilitated in several places by cutting the course of rivers and brooks straiter, particularly on the Smite, in the Vale of Belvoir.

Retford, Dec. 22, 1794.

SIR,

IT may be right to add to the drainage paragraph above, " The old drain was formed into land by sloping down the banks thereof, and throwing therein the earth that came out of the new drain."

The following remarks respecting the general drainage of this county, I leave you to dispose of in such manner as you think most proper. The drainage of every country greatly depends upon the state of the streams and rivers flowing through the same. There are many rivers flowing through this county, from west to east. Their average fall is from twenty inches to a foot in a mile, from their source to the tide's way. The rivers and main drains in the north, part of the county are inspected by a jury, called a water-jury, twenty-four in number, being freeholders of the county, possessing ten pounds per year of landed property, who make a survey of the rivers, &c. four times in a year. They make their report to a court called a court of sewers, which is holden at certain times of the year. I have frequently conversed with several of these jurymen respecting the business of their surveys; who say, they go upon that business because they are obliged so to do; that they do not understand the nature of rivers, and the best method of improving them—that they seldom see the rivers a mile together; they order a few sand-beds to be removed, especially at the tails of mills; and as they are appointed only for one year, they slip over it as easy as they can. The effect produced in the last twenty years makes good this assertion.

In other parts of the county the rivers are committed to the inspection of one person, which I have known some years ago to have been very little attended to, being persons not conversant in that business. Was the improvement of the rivers of this county to be committed to the direction of professional men, there might be more improvement made in a very few years to drainage of lands, and facilitating the discharge of the flood-waters, than will be effected by the water-juries and inspectors, as before-mentioned, in a century. Men of experience might be employed in improving and repairing rivers and drains in a country, as advantageously as surveyors of high-roads.

If any part of the above comes within the agricultural plan, they are much at your service, to dispose of as you think proper, hoping you will excuse the rude state you receive them in.

I am, Sir,
your most obedient servant,
RICHARD DIXON.

To John Holmes, Esq. Retford.

Covered Drains—have been made in various places in different manners.

At Halloughton near Southwell, Mr. Pogson, tenant to Sir Richard Sutton, has done upwards of two hundred acres, twenty-eight yards to the acre, as follows:—In meadow and pasture land, went two spade grafts, or two feet deep; then with another instrument of four inches wide, took out the soil of the drain made by the spade, twelve inches deeper; covered it with the sods first dug out, if the ground was found strong enough to admit of it, otherwise put in some black thorns sufficient to bear the sods; afterwards filled up the whole of the drain to the surface, with the soil taken out. This method has been pursued for several years past, and has perfectly answered. Expence per acre, including allowance for ale, &c. two shillings and sixpence.

At Norwood Park, Sir Richard Sutton did some several years ago, with the difference of using, instead of black thorn, ling or heath, which may perhaps be preferable; as it is known to be incorroptible for ages in water, and the sharp hard ends of black thorn may sometimes gall the earth, and make it moulder into the drain, which the other will not.

At Gonalston, in Thurgarton hundred, the Rev. Mr. Clark has done a good deal, in the following masterly manner: where there are boggy springy lands, at the foot of rising grounds, taking the lowest level, he pushes on a drain strait up to the hill, keeping it to the same level; so that it is deeper and deeper from the surface as he advances; then cuts a cross-drain at top, at the same depth, to intercept the springs. He then bores with an auger in both drains, but particularly in the cross one, at about every five yards distance; sometimes as deep as twenty-two feet from the surface of the ground. The springs boil up very strong to the bottom of the level, and run off. He then makes a wall of stones, set an edge on both sides, nine to twelve inches high, covering with flat stones; then lays broken stones and rushes to prevent the mould falling in, and fills in with earth.

Since the first publication of this report, the practice of covered-draining has been much extended. Mr. Richard Milward, of Lower Hexgrave, near Southwell, has done 1379 yards in a very masterly manner with stone; the drains three feet wide at top, one and a half at bottom; depth, according to the springs, from four feet to nine; waterway in the drains when laid, six inches by ten. By this draining, twelve acres of wet spungy ground, not worth more than five shillings an acre, have been made worth at least fifteen shillings.

Mr. Breedon, of Ruddington, has drained much in the same manner as Mr. Clark.

Mr. Cook, of Red Hill, in draining, has reared up sods in the drain, and covered with others; which he reckons better than to leave a shoulder, which is apt to moulder in.

In the coal land a good deal of covered draining is done, two and a half or three feet deep; *i. e.* two or three spits deep with a broad spade, then the bottom taken out with a narrow one, filling generally with small broken stones. Mr. Chambers, on similar land in Derbyshire, close adjoining, walls most of his drains, using no bottom spade. He makes the bottom two feet wide, then walls upon it, leaving six inches in width, and twelve in depth, for the water; and puts on a stone for a coverer, which he thinks a preferable method.

SECT. II.—WATERING.

THERE is certainly in this county a great opportunity of improvement by watering, as may be judged by the account of the running streams, inserted under the head of waters,

One hundred acres of water meadow have been made at Clumber Park, in a masterly stile, by Mr. Marson, the able manager of the Duke of Newcastle's improvements, according to the written and personal instructions of Mr. Boswell, of Piddletown, Dorsetshire.

On Whitmoor farm, near Thoresby, twelve acres can be watered with good effect, which were began by Mr. Samuel Sherring, agent to the late Duke of Kingston, at his own expence, (being tenant) except materials for the wears, &c finished by Mr. Pickin, agent to Lord Newark, (being the present tenant) also at his own cost. Mr. Pickin apprehends that Lord Newark has twenty acres more of boggy ground, capable of that improvement.

Robert Ramsden, Esq. of Carlton, near Worksop, employed a man from Gloucestershire to lay out a meadow for watering. It is well executed, and the whole of it might have answered as well as a small strip of it has done, every year since it was made, and upon which there was, last year, so much grass that two great crops of hay might have been cut from it. The part that was not flooded, for want of the command of water from another brook, was very inferior, both in quantity of hay and after-grass; also the bit that was watered is free from moss, rushes and other weeds, which impede the growth of the grass on the part not flooded. There are many other low grounds that would be greatly improved by properly watering them.

Mr. Brettell, tenant to Sir R. Sutton, at Thurgarton, has made a considerable number of acres of water-meadow, which seem to answer very well. Mr. Flinders, of Cathorp Mill, has also done several acres.

In the Coal District, intelligent persons doubt whether watering might not be prejudicial to the land, from the pernicious quality of the coal and iron with which the waters are impregnated; but a trial of this might be desirable on a small scale. An impediment often arises to this improvement, from an intermixture of property, from which it may be impossible to bring the water on, or carry it off again, without cutting through, or injuring the land of a neighbour. But it is to be hoped that mutual benefit, when understood, will induce contiguous proprietors to join in the work, and regulate their respective interests, as is seen in Wiltshire and Berkshire; where, through the whole course of a river, for many miles together, there is a continuation of these watered meadows, each person being served with water in turns. Another impediment arises from water courses not being sufficiently scoured out to make an outfall, which may be remedied by a due exertion of the law of sewers.

SECT. III.—MANURING, PARING AND BURNING.

MANURES.

Farm yard dung is, as in other places, universally used in preference to other manures, where a sufficient quantity of it can be procured.

In the Forest District.—The best is observed by Mr. Bower to be made by beasts fed with oil cakes.

Robert Ramsden, Esq. of Carlton, observes that it has been long a practice with strong soils, to plough in manure in the winter; but very few people have followed that method upon hot sandy soils: it however answers very well even upon such lands after a wheat crop, which is intended for a summer fallow with turnips, and which land is afterwards worked very much in the hottest weather to get out the twitch grass. This has been fully proved in a farm near Carlton, for several years past and particularly the last year, by part of a field which was so managed, upon which there was a much finer crop of turnips than upon the other part, which was managed in the general way, viz. by ploughing it in winter, without any manure, making a clean fallow in the hot weather in summer, and ploughing in the manure immediately before sowing the turnips.

In the Clay District—Farm yard dung is generally laid on the fallows, but good farmers wish to keep it for their grass grounds.

In the Vale of Belvoir—Farm yard dung is the principal manure; dunghills are made but not so carefully as might be. Mr. Pocklington has dressed with dung immediately after harvest: he must have a year's dung before hand for this.

In the Lime and Coal District—Farm yard dung twelve loads to the acre, is laid on the fallow.

LIME.

In the Forest District—Lime is almost universally used on the fallow for turnips, the quantity from one to two chalders, of thirty two strikes. Mr. Bower limes for the sake of his seeds, with one chalder of lime an acre in spring, besides farm yard manures, following it sometimes with fresh manure; he considers lime as hurtful without farm yard manure.

In the Trent Bank District—The lime used is chiefly from Newark, which is of a stronger nature than Kirkby or Mansfield Woodhouse lime; from having, as Mr. Sikes observes, more animal matter in it, the other more siliceous earth: from two to two chaldrons and a half per acre on the fallow for turnips. Mr. Sikes, who joins a knowledge of chymistry to that of agriculture, has made various experiments on manures as follows:—He thinks the good effects of lime much increased by mixing coal ashes with it immediately when drawn from the kiln, setting it on fire and letting it burn as long as it will; as by this process he concludes, the mixture approaches to the properties of the caustick alkali: the vegetable matter of the coal when calcined, producing a fixed alkaline salt which, when mixed with lime, forms a species of caustick alkali, analogous to soap boilers ashes, which have been supposed to fertilize in a considerable degree.

Mr. Sikes expects lime stone powder to do as well on grass land in a strong clay as lime, but has not tried it yet.

N.B. The gypsum used in Germany, is found to answer better raw than calcined. For grass land he would always mix lime with earth.

In the Clay District—Lime has been partially introduced for some years back, its effects being much disputed, and appearing indeed to be different in different places; owing perhaps to almost imperceptible differences of soil, or to the prior state of culture of the lands; the common proportion is twelve quarters per acre.

As to the effects of Lime on the Clay Soil, I have received the following information:

Mr. Calvert, of Darlton, on a cold clay soil, has laid from one to as far as twenty chaldrons of lime an acre, and found no benefit whatsoever. He used the Knottingly soapy lime from Yorkshire, which is much esteemed. He tried it for several years, having had two or three sloop loads. As he informed me, Mr. Cartwright of Marnham, by persisting in it, spoiled a close entirely.

Mr. Musgrave, of Kirklington, one year, on purpose for a trial, limed in Kirklington for his mistress, and in Halam for himself, two chaldrons, or sixty-four strikes an acre, some of Newark and some of forest lime; and in another part no lime, on a summer fallow red clayey loam. He dressed all with dung at the same time. He saw no difference in the crop: but where he laid the heaps of lime, nothing has grown since. His own land has never come about since that time. He apprehends many are drawn into lime by example only. If of any service, it is to lighten; but good fresh soil laid on is much better.

Mr. Brocksop, of Kirklington, manures his fallows with ten loads of dung, and six quarters of lime, with good success. On his first coming, he limed a piece of land with eight quarters of lime an acre without dung (laying only a very little on the worst parts), and sowed part wheat, part barley and clover. The barley was much better than on two lands which were left unlimed. The next year the beans, which followed the wheat, were a foot longer than where unlimed, and the clover remarkably good.

Lime is likewise used with good effect in Halloughton, Hockerton, and many other places; but from the above instances, the benefit of it does not seem to be fully established in clays.—Vid. Mr. Green's Observations, Appendix, No. X.

About Midsummer Mr. Cooke, of Easthorpe, in the parish of Southwell, puts ten quarters of lime upon an acre; he is sure it answers, by keeping the strong clay light, and can work it almost any time after the first ploughing. He hath tried it upon hazel land, but thinks it of most use upon clay. An experiment was made in a clay close, part without any manure at all, the rest limed; the lime brought much the best crop. He is of opinion that the lime is of full as much use the second year after it is laid on, as the first, and of service even the third. In Morton Field, it is customary either to lime the land and fold the sheep upon it, or use manure. Mr. Cooke saith there are much the best crops where liming and folding are practised. He can distinguish the difference a great way off. The Newark lime much the strongest. He hath used both.

Mr. Cooke comes from Long Eaton, in Derbyshire, where they lime both sand and clay; and he thinks it answers both equally.

In the Vale of Belvoir.—Lime is used in a small quantity, viz. five quarters an acre; by many not at all. Mr. Pocklington uses ten quarters. Great expectations are entertained of improvement from the Derbyshire Crich lime, being brought by the Cromford and Grantham canals (the latter of which runs through the whole of the vale,) when the navigation is complete. Mr. Pocklington and others have thoughts of bringing the raw lime stone and burning it; but as the carriage is paid by weight, query, whether this will answer.

In the Lime and Coal District—On limestone land, lime is laid sometimes ten quarters on fallow for wheat about August; for turnips the earlier the better.

The Rev. Dr. Coke, of Brookhill, near Mansfield, makes the following observation on the Derbyshire Crich lime.

" There has long been wanted in the husbandry of this country a species of manure which would answer as a top dressing, as well as when intermixed with the soil. This seems now to be acquired by the introduction of the Derbyshire lime. It is well known that in the common course of husbandry, our farms can only produce a certain proportion of manure; but by having recourse to this lime, the herbage is improved, and the vegetative quality of the soil is promoted. The stone from which this lime is produced is of a bluish colour and a hard nature. After it is deprived of its fixed air by calcination, it assumes the whiteness of chalk, which is a sure test of its being free from any mineral particles, and of its purity as a calcareous earth. There are immense rocks of this stone about Cromford and Crich, in Derbyshire, which may be easily transported by means of the Cromford Canal, to the principal parts of the county of Nottingham. The effect of this lime, when used, is so striking, that it may be seen to a small compass where it has been spread, and wherever a heap of it has been laid down, the grass is in greater abundance than in any other part of the field. It destroys the moss, and corrects that sourness so much complained of by the farmer in his grass land. The best method of using it is to set it in small heaps; water and spread it while it is quick. This is best done in the latter end of April or beginning of May. In consequence of this treatment, our pastures acquire a sweet herbage, abounding with white clover, and the best natural grasses. In a country where marl is not to be had, this proves a most valuable acquisition to its agriculture. It deserves likewise the same attention from the farmer in all his fallows, as it insures to him excellent crops of corn in consequence of its application.

In laying as lime it will I believe be found more advantageous to lay it altogether in a heap in the field, and water it as you throw it up, and cover it well with earth, you will find it go much further when you spread it on the land, and have a better effect on manure.

N.B. It is the Devonshire practice to mix the lime intimately with earth some months before it is spread, by ploughing the headland (there called the forehead) very deep all round the field, chopping it fine with mattocks, covering the lime with it, which is laid in a small ridge, quite round, and mixing the whole well together, as soon as it will slake with the weather.

On coal land lime as above, sometimes the Crich lime from Derbyshire is used, which is reckoned the best, sometimes the hungry lime of this country."

DOVE MANURE

Is not much used in the Forest District on the sand land. Mr. Birket, of Clumber, used it about four quarters an acre on turnips.

In the Trent Bank District,—four or five quarters an acre are laid on grass. Mr. Sikes lays it in a heap in a barn, and turns it till it falls.

In the Clay District—Dove manure is produced in great quantities, more pigeons being kept than are probably in any part of England. It is used as a top-dressing for wheat, at about three quarters per acre; but the greatest part of it is bought up at one shilling to fourteen-pence a strike, and carried up to the limestone part of the county, or into Derbyshire, where it is supposed to do more service.

It has been said, as an apology for the farmers in this district, suffering their dove manure to be carried away from them, that the money might be laid out in other sorts of manure, to more advantage for their land.

In the Vale of Belvoir—Dove manure is used by Mr. Pocklington, and some others, as a top dressing on barley, about ten strikes an acre, at one shilling.

In the Lime and Coal District—it is sometimes used on limestone land, as a top dressing on seeds, sometimes ploughed in very thin, about two quarters an acre. Mr. Chambers (more particularly mentioned under the head of Cultivation) thinks the latter the best.

It is little used *on the Coal Land*—being hard to be got, but Mr. Chambers supposes it the best manure of all if it could be had.

BONE DUST.

In the Forest District—Bone dust ground by mills on purpose, has been used by Mr. Birket, of Clumber, and Mr. Bower, of Drayton, to great advantage, as a top dressing for turnips; one chalder an acre, at one shilling and sixpence a strike besides carriage.

Mr. Wright, of Ranby, laid on for turnips, fifty bushels of bone dust an acre: the crop was very good; the seeds made no appearance the first year, owing perhaps to a dry summer; but the second, and every year after, were equal to where they were dunged.

In the Trent Bank District—it has been laid on grass, twenty strikes to an acre with great effect; the fresher it is the better.

RAPE DUST.

In the Forest District—Rape dust or oil cake from rape, has been tried by Mr. Flower, of Ollerton, on turnips; ten hundred weight an acre, at four pounds a ton. The turnips went off on the first sowing, but at the second, succeeded very well. Mr. Flower tried the following experiment: 1. Malt combs; 2. Ground bones; 3. Horn shavings; 4. Stable dung; 5. Salt; 6. Baron Van Haake's Powder, on strips of low grass land coming to a point. He found them succeed in this order: 1. Malt combs; 2. Stable dung; 3. and 4. Horn shavings and ground bones, equal 5. and 6. Salt and Van Haake, good for nothing.

GREEN MANURES.

In the Forest District—Mr. Wright has ploughed in clover with success, as a layer for wheat: has tried buckwheat, rolling it down when in flower and ploughing it in, but found no benefit from it.

MALT COMBS

Are used in the Forest District for seeds and tillage, six quarters an acre for turnips, at five shillings a quarter.

SCRAPINGS OF OILED LEATHER

Are used in the Forest District for hops at sixpence a bushel.

BOG EARTH.

Mr. Birket, of Clumber Park, has used sixty loads of black bog earth an acre, on seeds with good success. Had not the same on turnips.

GYPSUM OR PLAISTER.

In the Trent Bank District—Gypsum or plaister, the best of which is produced at Beacon-Hill, near Newark, has been tried by Mr. Sikes three years together, in the same manner and with the same bad success, as Sir Richard Sutton and Mr. Calvert had, as mentioned hereafter.*

In the Clay District—it was tried by Sir Richard Sutton for three years running, without success; but in case other persons should be inclined to make the experiment, it may be proper to mention that the quantity recommended by

---

* Vide Young's Annals of Agriculture, and vide Clay District, and No. VI. Appendix.

German writers, who speak highly of its effect, is about six strikes, ground fine, per acre, as a top dressing for any kind of corn; but particularly recommended for natural or artificial grasses.

Sir Richard Sutton has found great benefit to grass ground from skerry stone, found under the red loam, broken small, and laid on at the rate of five tons an acre.

WHALE BLUBBER.

Whale Blubber—mixed with soil from privies, has been used by Mr. Sikes, which had been previously mixed with lime. A hogshead of whale blubber, of sixteen hundred weight, to ten cart loads of soil, laid together for six months, and turned twice, laid on grass land where the grass was growing quick, three or four acres at a time; Mr. Sikes brought his sheep in again in a week.

SOOT.

About four quarters an acre are laid on wheat, which on cold land is sure to succed.

COMPOST DUNGHILL.

1. Mix one hundred loads of earth with ten chaldrons of lime, about May; let them lie together till the lime is fallen, but not run to mortar; then turn it over; lay seventy loads of stable dung in a heap close to it. When in high putrid heat, which will be perhaps in four months, lay a layer of this and a layer of earth, two thirds of manure to one of earth, and so go through the hill; turn it over in the spring, and lay it on in March or April: eight loads an acre on grass.

2. Mix lime and earth as before, and turn it; then cover it with soil from privies, and coal ashes, about one third in quantity; lay it at top for some months in an oblong heap, then turn and mix all together, letting it lie some months longer, and lay about eight loads an acre on grass; these two receipts are from Mr. Sikes, who adds that he finds road drift good upon clay land.

PARING AND BURNING.

In the Clay District—is sometimes used, but in the opinion of intelligent persons appears to be a dangerous practice, unless done very judiciously, and the land well supported with manure afterwards. Lands in Norwell lordship have been entirely spoiled by it.

In the Vale of Belvoir—Paring and burning is used sometimes in breaking up old turf on inferior land, for 1. turnips; 2. wheat or oats; 3. fallow. Mr. Pocklington observes, that land which eats bare, should never be burnt; only where it is rough.

In the Lime and Coal District—Paring and burning is used pretty much on limestone and coal land, where the land has lain long, and the sward gets very tough; though not so much on coal as limestone.

A tenant of the Duke of Portland's, at Bulwell Wood, is said to have ruined his natural grass by burning.

Mr. Green of Bankwood Farm, at Thurgarton, gives the following account of the effects of paring and burning:

" My manner of cultivating my land at Thurgarton, when I entered upon it in 1785, and for two or three years after, was different from that in which it had been treated before. Paring and burning had been the usual beginning, then two crops of barley, &c. which I had reason to think injured the land much. I began upon breaking up grass land to sow beans and pease, then wheat, but not finding the wheat crop answer, was obliged to try some other method. I was not willing to submit to the mode of paring, as, it being woodland cold land, it is in my opinion burning the best part of the land; which is as far as the grass roots go, and reducing the quantity of the soil: and after the two first years, I believe the quality. But being desirous to try some experiment first, I pared a small piece of land in a field, and adjoining to it ploughed up a small piece of grass land, and sowed both with turnips; but as the burnt dust was manure for the present crop, I thought in fairness the grass ploughed ought to have some, and laid on a smaller quantity, as about eight common cart loads an acre; but the grass land was ploughed only once before sowing (which I found to be an error in me.) The turnip crop was not so good as upon the pared land; but having the spring following sowed both parts with barley, the grass land ploughed brought the best crop, and so were the following crops till laid down with seeds. In the seed pasture I saw no difference. Secondly, in another field I ploughed a piece of land at the beginning of winter, by way of winter fallowing; and in the spring broke it to pieces in the best manner I could, laying about the same quantity of dung as abovementioned. Though in this method I found the land full of turf sods, I did not think or find it the worse for it. In the centre of the land was a part of it, with much stronger and rougher grass upon it. I pared that part and burned it, and sowed all the plot of land with rape seed for fattening sheep; but upon the winter fallowed I had much the best crop of rape, and it had likewise the best crop of grain afterwards, which convinces me that it is better not to burn the sod upon such sort of land. By ploughing it as beforementioned, I find the sod or turf manure for that land. The expence of paring and burning at a low price would be fifteen shillings per acre; and if I had not dung from my farm yard, without injuring or robbing other parts of my agriculture, *lime* will do, which I have tried and found to answer, laying so small a quantity as forty-eight strikes an acre, instead of the dung beforementioned, and treating the land in the same manner; and after the rape taking a barley crop, then an oat crop, afterwards a fallow, with rape or turnips; then sowing barley with seeds in this proportion, red clover, 8lb. white, ditto; one bushel and a half of hay-seeds, with a little ryegrass. This produces a pasture much to my satisfaction; and by pasturing the said land for three years, I find it in

I

a much improved state. Instead of fallowing when broke up again, I am quite satisfied it will answer for wheat to be sowed on the seed ground after lying the three years. I have made the trial and found it to answer. After the wheat I would sow oats, then fallow for rape as usual, and seed down as before: by this mode of management I find the land I occupy, which is woodland land, and nine years ago in a reduced state, to be much improved. By the use of seeds I keep considerably more stock, and by the improvement in my breed by the new Leicestershire sort, can keep more in number, make them fatter, and rise to greater weight in the same time than I could nine years ago.

Dr. E. Coke, of Brookhill, in the Lime and Coal District, says, " If land has lain any considerable time, and is full of bushes, we pare and burn it; if it has lain only a few years, we plough up the leys early in the spring, on which we sow oats. After the oats are carried off, we spread a quantity of lime on the stubble, and turn it down; in which state it lies till spring, when it is prepared by several ploughings for barley; along with which the red clover is sown: this is mown the year after, and wheat sown upon the clover leys; this practice is only used upon those loams that are dry, and upon a stoney bottom. After the wheat we have recourse to a fallow on which we sow turnips, or plant cabbage; then barley, clover, and wheat, and so to a fallow again. Some farmers throw in a crop of oats after the wheat, but this is wretched husbandry. On those loams that are moist, and have a clayey bottom, it is the constant practice to give them a complete summer fallow, during which time they are well limed, and then sown with wheat at Michaelmas. At the spring red clover is sown upon the wheat about May, and harrowed in with a bush harrow; this is mown the year after, and on the leys of it we sow wheat, or oats, the spring following. In some of our stiffer soils, during the fallow year, after manuring and liming, we plant cabbages on four feet ridges, which are succeeded by early oats, clover, and wheat. We keep ploughing between the ridges during the summer, which promotes the pulverization, and destroys the weeds. When we wish to lay our land down, we find no method more advantageous than after a summer fallow to sow wheat, and the seeds in the spring fallowing, as they are found to answer after this method better than any other. They should be harrowed in with a harrow drawn with thorns, to prevent them being let in too deep.

In regard to paring and burning, it is frequently practised by these farmers, but is always a symptom of bad husbandry; because a good farmer will never permit his land to lie in that neglected state, until it wants paring and burning. The general mode, after paring and burning the turf, is to spread the ashes, and along with them a quantity of lime. These are turned down with one ploughing, upon which turnips are sown, on the dry land, and wheat upon that which is moist. I have found by an experiment which I have made upon stiff land, that it is better after the lime and ashes are turned down, not to sow upon one plouging, but to crosscut and plough several times, during the summer; by that means pulverizing the soil, and mixing the ashes and lime intimately with it, and thus prepare it for wheat at Michaelmas."

Mr. Calvert, in a letter to Sir R. Sutton, says, " Ashes may be a good management as far as they go, and while their effects continue; but though I preserve what I can of that article, and approve the use, *I must confess myself an enemy to burning swarth*, (what has been too much the practice in this and other counties) unless the soil be remarkably thick indeed. The ashes remaining, after the richest part of the sod (were it reduced to manure by putrefaction) has escaped in smoke, contain a much less bulk than the soil itself, and though productive of one good crop, and perhaps more, the effects soon disappear; and in many instances I have known a barrenness ensue, which a long series of years has not been sufficient to remedy, though much expence and pains have been bestowed thereon. I believe the damage is often encreased by the neglect of persons who burn the sods, in letting the fires rise to such a height that the earth contained in the swarth is burnt to the nature of a brick, and when that is the case I suppose that neither time nor art will ever bring it back to earth again; for I have observed fields ploughed up, in which the burnt earth is as discoverable as if fresh done, though the land has not been burnt in the memory of man. This kind of doctrine will not be relished by the favourers of paring and burning, as they are well aware it is the surest method to obtain an excellent crop; and I must confess I do not recollect I ever knew an instance wherein it has failed. But notwithstanding the farmer may advance many arguments to prove its utility in production of crops, yet if the landholder must abate as much in the real value of his estate, it must tend at last to a degeneration of the land so abused; and, consequently, a bane to future occupation."

SECT. IV.—WEEDING.

Weeding is practised in this county in the usual way by the hand hoe or weeding hook, but is in general not sufficiently attended to; particularly in the clays, where very foul crops may generally be seen. There seems to be no idea of weeding beans by sheeping, *i. e.* turning sheep into them as practised in Hertfordshire, Bedfordshire, and other parts. It is indeed very much the practice here to sow pease with the beans, which would preclude this way of weeding, as the sheep would eat the pease though not the beans. It is pretended there is an advantage in sowing them mixed, viz. the smothering the weeds, but I have observed these crops to be generally very full of weeds.

Some persons use the cultivator, described by Mr. Bower, or some very similar to it, to get out the twitch grass, or couch, as called in some parts of England.

Mr. Wright, of Ranby, in the Forest District, to destroy bracken or fern, a very troublesome weed there, uses the following method: He has at the end of a stick, a blade with dull edges; a woman uses this to strike the stems and bruise them, and will do several acres in a day; this is repeated two or three times in a summer; the next morning a gummy consistence is found to exsude, and the bracken gradually disappears.

Extract of a Letter from Mr. CALVERT *to Sir* R. SUTTON.

"I think I have before hinted to you the great advantage I once acquired, (about three or four years ago) by hand-weeding a crop of wheat in the month of April. It was perfectly to appearance smothered by weeds, particularly what we call hariff, or herrif, so as that very little wheat was to be seen. I ordered my man to harrow it, which he did, till I perceived the harrows drew up more corn than I wished, and left very much weeds; the soil was a light silted land. He then desisted, and I employed a number of women to creep over it, with directions to pluck up every weed, if possible, however small; they obeyed my directions pretty well, which cost me five shillings per acre; and notwithstanding there scarcely appeared any wheat left after the operation, in about three weeks the crop looked well, and became sufficiently productive, yielding nearly four quarters per acre, and the wheat weighed sixty-five pounds per Winchester bushel."

SECT. V.—PLANTING.

There are large tracts of forest land in rabbit warrens, or sheep walks, which are of so poor a nature, that it is doubtful how they would answer to be taken in for husbandry; but there is little doubt that they would be of more value if planted than in their present state. For dry high ground, I should advise plantations of fir, oak, Spanish chesnut, larch, beech, and birch, taking out the fir, beech, and birch, where they have nursed up the others. Ash do not do well on the sand of the forest.

For low bottoms (if not too wet) I should from my own experience, recommend the red Huntingdonshire willow.

ON THE CULTIVATION OF WILLOWS.

The following observations relative to the method I make use of, in cultivating willows on waste moist lands, I flatter myself will not prove unacceptable to the Board of Agriculture. I have found, from experience, the advantage of it, and am convinced my country will be benefited, should it be generally adopted.

I would first advise the laying out the ground into lands, like hop lands, viz. from three to four yards wide, with a ditch on each side; three feet wide at the top; one foot at the bottom, and two and a half deep. The earth that comes out of the ditch should be thrown on the land. But if there is not full sufficient fall for the water to get off, the ditch should be deeper and wider, till you have near a yard of earth above the level of the water.

As soon as this is done, the ground must be double dug, viz. trenched two spades depth, except your ground be very boggy, which will afford room for the plants to shoot, and will save the expence of weeding, which otherwise must be incurred in the first summer after the plants are set; for if they are not kept clear of weeds the first year, the hopes of the planter will certainly be destroyed.

The willow I recommend as most advantageous on every account—is the broad-leaved red hearted Huntingdonshire willow; every other species I have tried, and find reason to give a decided preference to this.

The sets, or truncheons, may be cut from twenty inches to two feet long; particular care should be taken in the cutting, that the bark should not be fridged or bruised, or in any other respect injured; for in that case the plant will be weak and puny. They should be cut not on a block, but in the hand; obliquely and with a very sharp bill, or instrument. They must be dibbled into the earth by an iron crow, to the depth of fourteen or twenty inches, so that not more than six, or less than four appear above. If the truncheon should not fill the hole, the earth must be trampled close round it, in order that the air may be excluded. Care must be taken that the plant be set as the pole grows. The cuttings should be from poles of about three years growth. Maiden poles are the best; they should be set three foot asunder in the quincunx form, as thus:

* * * * *
 * * * * * *
* * * * *

Those truncheons will shoot out many branches, two or three of which will grow to poles if the land is good; if not, only one. Those poles I have sold at eight years growth, for 214l. per acre, neat money; the kids or brushwood pay for the felling. Had I suffered them to have stood two years longer, they would have produced 300l. per acre. Should any of the plants look weak or puny, or not shoot vigorously, it will be necessary to dig in a skuttle full of manure, to the roots, which I have no doubt in saying will pay.

Though I have planted no less than ten acres, I cannot say positively, from my own knowledge, what the value would have been had they remained on the ground for fifteen or twenty years, having been called on for sets by the gentlemen of the neighbourhood, which I have sold for three pound a thousand. I must here observe, that the stools from whence the sets are cut, shoot very luxuriously, and will produce from three to four poles.

The length of poles, at eight years growth, were from thirty-three to thirty-six feet, and most of them were large enough to make three rails, two at the bottom and one at the top; but the great use to which they are applied, is the purpose of making hurdles, flakes, gates, and other farming implements, being a wood uncommonly tough and light, owing, as I conceive, to a new method I made use of in planting them close to the ground. If it is the design of the planter to let them grow into timber (which I would venture to say would be far superior to *deal* for the purpose of flooring, or other light work, particularly as it will neither splinter nor fire; and if suffered to remain for twenty or twenty-five years, would make good masts for small craft, as they shoot up perfectly strait, and without any collateral branches) it is necessary, at the first or second year's growth, to observe which pole is the strongest, as the remaining poles must be cut away. In about fifteen years time I am led to suppose they will want thinning; of course the inferior must be taken out and the superior be suffered to remain.

The times of planting must be from January to the end of March; but the sets for that purpose should be cut from December to the end of February; when the sap is down.* If however there are people so injudicious as to sell sets in spring, it will be to the advantage of the purchaser to plant them, as the sap is then in the poles. The reason why many are induced to cut at that time, is on the supposed account of their pealing better; but I can affirm from experience, that poles cut in December, January, or February, and laid in rows upon the ground, or the ends put in water, will peal as well in the spring as at the usual time.

In regard to fencing, the planter should pay the greatest attention to it, otherwise his time and expence will be fruitless.

* And the reason is, that if poles are cut in the spring (the sap being up) the stool will at last be weakened by bleeding, if not killed; and of course prevented from shooting so vigorously as if cut at the preceding time.

ON INCLOSURES.

In regard to fencing any new inclosure, I should recommend (though it has been untried) the following method: First of all, to set the quicks, which should be young ones or seedlings, or otherwise hawes sown, the ground being first dug, but to be secured by a four or five sod bank on each side;—by which means the quicks will be preserved and attain more moisture; and, of course, being planted in the natural soil, will get into a hedge many years sooner; and if the planting of hedge row timber is an object, I am confident it will not injure the land. Those should be planted about four or five inches from each other, and from about one or two feet from the quicks, by which means the same fencing that protects the quicks will protect the trees; so that no additional expence will be incurred in regard to hedge-row trees injuring the land. I am convinced from many years observation, that it is not the case; for I have observed, that the warmth and shade it affords in summer and winter is a benefit rather than an injury. I have seen that the grass grows stronger, and that the cattle, by being kept dry and warm by the trees when they get up, do better. I should recommend oaks in preference to any other trees; as hedge-row oaks make the best timber, and injure the land less, on account of the tap root; and next to oaks, elms, but that species of them which does not shoot out suckers, both being of the greatest consequence to ship building.

The oaks may be planted when seedlings, or sown from acorns; but the elms should be planted from layers not more than three feet high: if seedling oaks are planted, I should recommend an acorn to be dibbled in between each plant, from four to five inches asunder; and when they begin to interfere with each other, which will take place when they are about the size of a good hedge-stake, then every other should be sawed out: I would except, how-

however, any remarkable fine plant. In about twelve or thirteen years, when they obtain a size sufficient for the purpose of making rails, gates, &c. the same method should be observed, and so on till they become timber trees.

Hedge-row Planting.—Is to be lamented, that in the new inclosures very little attention should have been paid to raising hedge-row timber, which is done at first with no more expence of fencing than the raising of the quick. There is, I believe, a general prejudice against trees in hedges, as being supposs'd to injure the land. Whatever may be the case, with regard to corn, which I apprehend to be much exaggerated, it does not appear to do any hurt to grass; and there cannot, in my opinion, be a better or more convenient method, for many purposes, as I have beforementioned, than always to leave a head-land in grass, as may be seen in many parts of Hertfordshire, under the name of hedge-greens. It is obvious, that a great quantity of timber may be raised in hedge-rows, which is better for various purposes, than what is raised in woods, particularly compass timber, as knees, crooks, &c. the most valuable pieces in ship building.

I believe I may venture to affirm, that trees so planted are likely in twenty-five to forty years, to equal the value of the land, whilst 'n their growth they have taken up no land from other purposes.

I have above (vide Inclosures) ventured to suggest a method of planting them to more advantage, and to raise a greater quantity than I have known practised.

CHAPTER XIII.

Live Stock.

SECTION I.

CATTLE.

FEW are reared in the Forest District. For feeding, after trying various sorts, Mr. Wright prefers the good sort of Irish cattle. He would not buy any that would not feed to fifty or sixty stone, of fourteen pounds.

Not many black cattle are bred in *the Trent Bank District*. On the Soar bank the breed of cows used to be indifferent. Of late they are got into a pretty good long-horned breed. They rear almost all their female calves; which, when young, are pastured amongst the sheep, and at three years old are taken into the dairy, and the old ones fed off.

Some persons have improved their stock by the Dishley breed. Mr. Breedon, of Ruddington; Mr. Bettison, of Holme Pierrepont; with three others, bought the Garrick bull, at Mr. Fowler of Rollwright's sale, for two hundred and five guineas. The bull called Young Garrick was bought by Mr. Rowland, of Stamford, in September last, for one hundred and fifty guineas. A bull was bought at the sale of Mr. Paget, of Ibstock, in Leicestershire, for four hundred guineas; of which Mr. Sandy, of Holme Pierrepont, had a sixth part.

Mr. Bettison, of Holme Pierrepont, observes that much improvement may be made in the breed of beasts and black horses in this part of the county. For such as are the occupiers of small farms, and desirous to improve their stock, and not able, the most obvious mode presents itself for the landlord to form a committee out of the most intelligible class of tenants, who shall procure either by *hire* or *purchase*, such and so many male stock, of different sorts, as shall in their judgment, be most proper for the improvement of different breeds. The landlord to be answerable in the first instance, for such *hiring* or purchase, and the tenant in proportion to the quantities of their respective stocks, contribute so much in return annually) to the landlord, and according to chances in general to reimburse him with interest.

The beasts reared in *the Clay District* are generally of a poor coarse kind, commonly called wood land beasts. Some gentlemen and principal farmers are endeavouring to introduce a better sort.

Peter Pegge Burnel, Esq. of Winkborn, a gentleman very understanding in husbandry, keeps the Yorkshire short horned breed, and computes that they are worth at least, as much at three years old, as the old breed at four. Mr. Turnell of Stokeham rears the same.

In the Vale of Belvoir—A good many beasts are still reared, though that business is on the decline; a mixture of the long and short horned breed which generally wants improvement.

In the Lime and Coal District—The black cattle are very indifferent; a mixture of long and short horned woodland beasts.

SECT. II.—SHEEP.

In the Forest District.

The old Forest breed are a small polled breed (though some are horned) with grey faces and legs; the fleeces of which may run from thirteen to eighteen to the tod of twenty-eight pounds; the wool fine, the price of 1792 being from thirty-four to thirty-six shillings; that of 1793, from twenty-nine to thirty-one shillings. The carcases fat, from seven to nine pounds a quarter. In the inclosed farms, the breed has been much improved of late years, by various crosses; sometimes the Lincolnshire pasture sort, but of late more the new Leicestershire, or Dishley. Mr. Birket, of Clumber, crosses the new Leicestershire, or Dishley, with the forest kind; reckons their wool seven and eight to the tod; the price sometimes two-thirds, sometimes three-fourths, of the true forest wools; the carcases eighteen to twenty pounds the quarter. The breed in Thoresby Park are likewise the forest breed, crossed with the Dishley or new Leicestershire; the tups, coming from Mr. Bettison of Holme Pierrepont. They are a round compact kind; the fleeces from six to eight to the tod, or, at an average, about four pounds a fleece; the carcases from seventeen to twenty-two pounds a quarter. The following instance was given me of the improvement from this cross:

Mr. Jones, of Arnold, used to sell his forest wethers, at four years old, for fourteen shillings; now on inclosed land, *not yet broken up*, shearlings, *i. e.* under two years old, of Mr. Bakewell's breed, with forest ewes, turniped not more than four months, at thirty-four shillings.

Mr. Flower tried the experiment of feeding four different breeds together; kept whilst getting ready for market, in the same manner on good seeds and turnips. 1. All Lincolnshire of the middle breed sold in August at three and a half years old, for twenty-eight shillings a piece. 2. Crossed between a forest tup and Lincolnshire ewes of the same age, thirty shillings. 3. All forest of same age, nearly as all Lincolnshire, twenty-eight shillings. 4. New Leicester, at two years and one month, forty-six shillings.

	lb.
Fleece of all Lincolnshire weighed	10
Forest and Lincolnshire mixed	8
All forest	5
New Leicester	.7

Mr. Wright breeds from large Northumberland ewes, and new Leicestershire tups. In 1792, he made upwards of thirty-nine shillings a head of his wethers, at thirteen or fourteen months old, reckoning in the lambs wool; he now thinks it better not to clip lambs. His ewes tod four or five; his hogs, when not clipped as lambs, four or less, at twenty-eight pounds to the tod; carcase at a year old, (at which age he always sells) eighteen pounds a quarter.

Mr. Bower breeds in the same manner as Mr. Wright; sells off all shearlings, which run four to the tod, carcase eighteen pounds a quarter. The wool, this year, about two-thirds of the value of forest wool, but in general three fourths. For the last three years he got twenty-five shillings, now but eighteen shillings.

Mr. Wright is now of opinion that if the forest breed were gradually improved by Leicestershire tups, and the native forest ewes, it would answer better than having either the Northumberland or Leicestershire ewes out of those counties; removing them from rich soil to poor ewes may be one reason why there is sometimes great loss in the lambs from a disorder provincially called the Rickets.

Mr. Ramsden, of Carlton, in Lindrick, has some time since some of the Southdown breed. Sir Richard Sutton last year procured twenty-four ewes and a tup from the Southdowns, and has kept his Heder lambs for tups, in case any person in the neighbourhood should wish to breed from them, but has had no demand for them. There seems to be no idea in this county of improvement in the breed of sheep but from the New Leicestershire sort. On the other hand, I am informed that some persons in the part of Lincolnshire, near Newark, where for some years past they have been crossing with New Leicestershire, finding their stock deficient in weight of carcase and fleece, are returning to the Old Lincolnshire breed.

In the Trent Bank District—The sheep have been much improved for many years past, by tups of the Lincolnshire and new Leicestershire sort; but of late many more of the latter. It is become of late a principal object of attention, and many breeders are spreading the improvement, by letting out their tups, some at as high a price as one hundred guineas.* Amongst others, Mr. Breedon of Ruddington; Messrs. Stubbings and Bettison, of Holme Pierrepont; Mr. Deverell, of Clifton; Mr. Buckley, of Normanton Hill; and Mr. Maltby, of Hoveringham, are noted. Mr. Breedon was one of the first who began, about twenty-four years ago. He had his first tup from Mr. Bakewell, and except two years, has had one or more from him every year since. He breeds in and in, to keep up the kind; reckons his shearlings, or two years old sheep, fat, from twenty to twenty-eight pounds a quarter; his three years old, from twenty-five to thirty-five pounds a quarter. Their wool, under four to the tod, the value of it about two-thirds of the forest sort. These sheep are feed a good deal on artificial grasses; red clover mixed with some white. Of late the white is rather preferred, as accidents happen from the red being too luxuriant, causing an overflowing of the gall.

Mr. Breedon thinks it may be an improvement, to clip the lamb hogs, and sell them off that summer, at one year and some months old, getting rid of them a year sooner than usual. Mr. Sikes has a fine breed of sheep, having for these last fifteen years had all Dishley tups. He observes, the Lincolnshire are desirable for quantity of wool, the Leicestershire for mould. Mr. Sikes always keeps salted hay; has standing racks for his sheep, which he fills the beginning of September, and keeps on till Christmas: two pecks of salt are used to a load. He once led his clover, supposed to be quite spoiled with wet, and salted it. He put one hundred and twenty hogs (*i. e.* lambs

---

* Mr. Bakewell has let this year two theaveling tups, each sheep between two people, at four hundred guineas each.

after five or six months) to turnips, with his hay in racks, in a very wet season. He did not lose one by the water: They eat every morsel of it. Horses are well kept on it.

Mr. Bettison, of Holme Pierrepont, says—"the species of stock upon farms are principally sheep and beasts. The sheep are of the polled kind, are in general descended from those bred by Mr. Bakewell, of Dishley, in Leicestershire, whose attention and study has been to obtain and produce the best and completest formed animal of the sheep kind; and circulating those sheep to other breeders, who have, by experience, proved that they make a more profitable return for what they consume than any other sort;—and not only that, but suit more various kinds of land. To enter into any investigation of the causes of which I have here advanced, would be perhaps thought a little out of the line of the present inquiries, and would, in some degree, be deemed curiosity and conjecture; it may be enough to say at *present*, that the effects I have mentioned are proved to demonstration; and however wild and fanciful many have been supposed, for giving large sums of money for the use of rams, for the purpose of obtaining this breed of sheep, it is now generally agreed, that those who have risked such sums, have received the most profit in proportion; and it is recommended to those who doubt this conclusion, to bestow a little dispassionate and serious observation upon the character, conduct, and stock of those men, who have for many years back been exerting their utmost endeavours, by incessant study and attention, to acquire the sort of sheep I have above mentioned; and instead of meeting with men actuated by *whim* and *caprice* as hath been *frequently alledged* by those who stop the avenues to reason by *prejudice*, I trust they will find men who are actuated by a desire of real profit, and have given the high prices I have mentioned from the motive which all men of business ought to keep in view, that of employing their capital in the most advantageous way."

For the sheep kept by Mr. Nevile, in the tongue of land east of Trent, vide his letter ante.

In the Clay District.—The fallow sheep are a poor breed; a mixture generally between the forest and Lincolnshire pasture sheep. In the inclosures many farmers have raised their breed, by getting more into the Lincolnshire, and of late into the new Leicestershire sort; particularly in Thurgarton Hundred, adjoining to the Trent Bank country. Mr. Turnell, of Stokeham, breeds in the same manner as Mr. Wright in the forest, viz. large Northumberland ewes with Dishley tups.

In the Vale of Belvoir.—the sheep are much improved of late by the Leicestershire cross. Their wool three and four to the tod.—Good shearlings twenty-four to twenty-eight pounds the quarter; common ones, from sixteen or eighteen to twenty-three.—Wool three and four to the tod.

In the Lime and Coal District.—As to sheep, an improved breed has not been much attended to.

On the Limestone, pretty good ewes, fat, weigh from fifteen to sixteen pounds a quarter; wethers, (if not mixed with the forest breed) up to twenty pounds; fleeces, about seven to the tod; wool inferior to the forest; last year about twenty-four shillings. The sheep formerly consisted of the small forest, and the large Gritstone breed, which is now giving way to that of the Leicestershire. The coal land is much subject to rot sheep; lime stone much sounder.

Mr. Chambers has known hundreds cured of themselves on limestone land not eaten too bare; their livers healed again.*

* This information is very valuable, as it leads to a discovery of what may possibly prove the cure of this disease in other counties. Water impregnated with the fixed acid of lime in proper quantities, with change of pasture as soon as the disease appeared, might remove it.—*Mr. W. Fas.*

SECT. III.—HORSES AND THEIR USE IN HUSBANDRY COMPARED TO OXEN.

In the Trent Bank District—Some horses are bred, chiefly a middling kind of black cart horse, though the breed begins to be improved by Leicestershire stallions.

In the Clay District—Most farmers raise a foal or two every year, but of a middling kind of black cart horse, which calls for improvement.

Dr. Coke, of Brookhill, in the Lime and Coal District, says, " The breed of black horses is much attended to, and a great number of them are sold to the southern dealers, who come down to buy them."

Horses —The business of agriculture in this county is almost universally done by horses: those generally made use of, are a middling sort of black cart horses. Such fine teams are not seen here as in many of the southern counties. Mr. Jones, who rents the great tythe farm at Arnold, about 800 acres, performs all his work with nag horses, which he finds to work with more expedition on light land. It is become, within these few years, the general custom in the sand land, and begins to be so in the clays, to do all the latter orders with two horses a-breast, without a driver.

Oxen—are so little used, as scarce to make an exception. Mr. Bower, of West Drayton, however cultivates a forest farm of 100 acres entirely with four oxen, and is very well satisfied with them; they work wholly on straw and grass, and do an acre a day in winter, and five roods in the spring. They are used two in the morning, and two in the afternoon. The Duke of Newcastle employs some beasts in Clumber Park, particularly in getting out his wood. Sir Gervase Clifton has a team of Devonshire oxen. Sir Richard Sutton, Mr. Stubbings, of Holme Pierrepont, and Mr. Wilson, of Shelford Manor, employ some, and perhaps some few other persons who have not come to my knowledge.

SECT. IV.—HOGS

Are no great object in this county except for home use; no hams, bacon, or pork being sent out of it to my knowledge. The breed for bacon is the old lopped-eared sort. For pork, the Chinese dunky, or swing tailed sort. A mixture with the old sort has been much introduced.

SECT. V.—RABBETS.

Rabbet Warrens.—There were formerly many rabbet warrens in the Forest District. Those at Farnsfield, Clumber Park, Beakwood Park, Sanson Wood, and Haywood Oaks, have been destroyed. The following remain: Clipston, Peasefield, Inkersall, Oxton, Blidworth, Calverton, and Newsted. The land of some parts is so bad, that it is not likely to answer if taken up for husbandry. Some of it indeed has been tried and thrown up again.

SECT. VI.—POULTRY

Has never been made an object of particular attention. Few turkies are bred. Fowls are commonly of a bad breed, generally the game sort; which are raised as much for the diversion of cock-fighting as the table. Geese are reared for home use only, or the neighbouring markets, not to be sent away in droves, as from many northern counties.

SECT. VII.—PIGEONS.

In the Clay District—more pigeons are kept than are probably in any part of England. It is a well attested fact, that some years since, seven hundred dozen were sold on one market day at Tuxford, to a higler from Huntingdonshire, at the price of sixty-three pounds, or guineas.

SECT. VIII.—BEES

Are very little attended to in this county. Indeed the climate of England seems very ill adapted to them, from its variableness. The bees are tempted out to their destruction by the fine warm days, which we often have in winter. Of particular persons who have applied themselves to the keeping of bees, I have observed the stocks, of late years, to have been much diminished, perhaps from above twenty to under ten.

CHAP. XIV.

Rural Occonomy.

SECTION I.

LABOUR, SERVANTS, LABOURERS, HOURS OF LABOUR.

THE prices of labour are so various in different parts of the county, that nothing satisfactory is to be said on the subject. Within these few years, day labour is raised from one shilling to sixteen and eighteen pence a day; and, for the three harvest months, to two shillings. In harvest they expect likewise some beer. Task work is raised in proportion. Threshing is now for wheat, four shillings per quarter; barley, two shillings and sixpence; oats, one shilling and sixpence; which were, a year ago, three shillings, two shillings, and one shilling. The hours of labour are the common ones. The ploughing is generally done at one stretch. Boarding labourers in harvest, as is done in some counties, is not usual here.

SECT. II.—PROVISIONS.

Provisions are here much on a level with the neighbouring counties. They have been for a year or two past enormously high; beef and mutton sixpence, and even sixpence halfpenny per pound: but are now (in the latter end of 1797, and beginning of 1798) got down to fourpence, and fourpence halfpenny; and pork in the same proportion.

Butter has been lately from ten-pence to a shilling, and fourteen-pence; cheese, forty-four shillings, and forty-five shillings per hundred weight; bacon, eleven pence to one shilling a pound.

SECT. III.—FUEL.

The fuel used in this county is, almost, universally coal. A good deal is produced in the Lime and Coal District; and a good deal brought out of Derbyshire, by the Erwash, Cromford, and Chesterfield canals. Some is brought by water from Yorkshire, from the river Air; and some is distributed, by land carriage, from the pits. It is observable that since the canals have been opened, coal is become much dearer to all places, within a certain distance of the pits, the price having been greatly raised at the pits themselves. At Southwell, those that used to be laid down at ten shillings, and ten shillings and sixpence a ton, are now at fourteen shillings, and fifteen shillings.

At Mansfield and Worksop, coals are also risen. On the other hand, at Newark and Retford, and other places at a good distance from them, they are fallen.

In general, this county may be said to be supplied with fuel at a reasonable rate, which is a great advantage to the manufacturers, and the poor.

CHAPTER XV.

Political Oeconomy as connected with or affecting Agriculture.

SECTION I.

ROADS.

THE roads of this county are of late years much improved; many parishes having learned, from the example of the turnpikes, to form them properly, and have them executed under an understanding surveyor. Gentlemen and considerable farmers having taken on them the office of surveyor, has also contributed to their improvement. They remain, however, bad in many places in the Clays, and particularly in the Coal District, where there is a great deal of heavy carriage. In the Forest District too much is sometimes left to nature, where a little expence would make them perfectly good. The most approved system of making new roads on clay or wet bottoms, is, first throwing the soil from the sides, leaving a groove in the middle for the materials, beginning with kid or brush wood plentifully, then stones and gravel: if the gravel is very sharp and good, there is no occasion to round the road. Even a concave surface is found to answer very well; but where the materials are tender, it may be better to round it a little, but not so much as is frequently done, which is often dangerous, and hurtful to the road, by obliging carriages to keep one track.

SECT. II.—RIVER NAVIGATION AND CANALS.

There is a great trade carried on in this county by water, by means of the river Trent, and the different canals.

By the Trent are carried—

DOWNWARDS.

Lead, copper, coals, salt, from Cheshire, cheese, Staffordshire ware, corn, &c.

UPWARDS.

Raff or Norway timber, hemp, flax, iron, groceries, malt, corn, flints from Northfleet, near Gravesend, for the Staffordshire potteries.

By the Canal from Chesterfield, to Worksop and Retford, and to the Trent at Stockwith—

DOWNWARDS.

Coal, lead, Steetley stone, lime, and lime stone, chirt-stone, for the glass manufactories, coarse earthen ware, cast metal goods, and pig metal, oak timber and bark, and sail cloth.

UPWARDS.

Fir timber and deals, grain, malt, and flour, groceries, bar iron, and Cumberland ore, wines, spirits, and porter, hemp and flax, cotton-wool and yarn, Westmorland slate, and various sorts of small package.

UPWARDS AND DOWNWARDS.

Bricks, tiles, hops, and candle-wicks; other articles, however, bear but a small proportion to the coal, downwards; and the corn, groceries, foreign timber, and iron, upwards.

By the Erwash and Nottingham Canals—

DOWNWARDS.

Coals from the Nottinghamshire and Derbyshire pits.

UPWARDS.

Corn and malt, for the consumption of the country at the head of these navigations, which is very populous, are carried up by the Erwash canal, and are likely to be so by the Nottingham, when completed.

Great advantage is expected from their junction with the Cromford canal, in bringing lime from Crich, and other places in Derbyshire.

N.B. This is now compleated, and found to answer, particularly with regard to lime.

SECT. III.—FAIRS AND MARKETS.

The principal fairs are

At Nottingham 3.—1. March 7.—2. April 2. (moveable)[*] 3. Oct. 2. called goose fair, particularly for cheese. All of them for cattle and horses.

At Newark six.—1. on Friday before Careing Sunday, or Sunday fortnight before Easter.—2. May 14, or day

[*] Query—If Monday after Palm Sunday.

after, if a Sunday.—3. Whitsun Tuesday.—4. Lammas, or Maudlin Fair, on August 2, or if a Sunday, the day after.—5. All Saints Fair, 1st November, or if a Sunday, the day after.—6. St. Andrew's, Monday before December 11.—Great fairs for cattle, sheep, and horses.

At Mansfield.—April 5.—July 10.—2nd Thursday in October.—Horses, beasts, and cheese.

Worsop—May 21.—November 17.

Edwinstow.—October 24.—Both these chiefly for pigs and sheep.

Worksop.—March 31.—October 15.—Chiefly for beasts.

At Tuxford, for hops, September 25.

At Retford, for ditto, October 2.

MARKETS.

Nottingham	Saturday.
Newark	Wednesday.
Mansfield	Thursday.
Bingham	Thursday.
Worksop	Wednesday.
Tuxford	Monday.
Ollerton	Friday.
Southwell	Saturday.

SECT. IV.—COMMERCE AND MANUFACTURES.

The malting business is carried on to a great extent in this county, particularly at Nottingham, Newark, and Mansfield, and in many other places. A great deal of malt is sent up by the Trent and the canals, into Derbyshire, Cheshire and Lancashire. At Newark are great breweries, which vie with Burton upon Trent, in the trade to the Baltick and other parts. At Nottingham is a brewery, and another going to be established on an extensive plan.

OF NOTTINGHAMSHIRE. 139

The Stocking Trade—is the most anciently established manufacture in this county; the frame for knitting stockings having, it is said, been invented by one Lee, of Calverton.* It occupies a great many hands at Nottingham, and the villages for some miles round; as also at Mansfield, Southwell, and other places in its neighbourhood. Many new works of different kinds have been lately erected, as follows: Many cotton mills worked by water, to prepare the thread for the Manchester manufacture, and for stockings, and other purposes, as at Gamston, Lowdham, Papplewick, Southwell, Newark, Fiskerton, Mansfield, and Basford.†

At Cukney is a mill for combing wool, and another for spinning worsted, and one for polishing marble. At Arnold is a large woollen mill for both the former purposes: at Retford is a mill for combing woollen. These two are worked by steam. At Nottingham silk mills worked by horses. At Mansfield is a great trade in stone. Artificial marble is likewise made, and a considerable thread manufacture carried on, as also of British lace. At Nottingham is a white lead work, a foundery for making cast iron ware out of the pigs brought from Colebrook Dale, a dying and bleaching trade, and a manufacture of British lace by frame work. At Sutton, in Ashfield, a considerable pottery of coarse red ware, for garden pots, &c. At Upton, near Southwell, is a starch manufactory. At Retford, a sail-cloth manufactory.

* Thoroton, in his Antiquities, says, page 207, that, " At Calverton was born William Lee, master of arts in Cambridge, and heir to a pretty freehold here, who, seeing a woman knit, invented a loom to knit, in which he or his brother James performed and exercised before Queen Elisabeth; and leaving it to —— Aston, his apprentice, went beyond the seas, and was thereby esteemed the author of that ingenious engine, wherewith they now weave silk and other stockings, &c. This —— Aston added something to his master's invention; he was some time a miller at Thoroton, nigh which place he was born."

† Vide Appendix, No. XI.

SECT. V.—POOR.

Having before spoken of the poor's rates, I have only to add, that there are few counties in England where they will be found better lodged, cloathed or fed, or better provided with fuel. Most cottages have a garden, and potatoe garth, and few of them are without a web of cloth of their own spinning: many of them, particularly in the Clay District, have a few acres of land annexed to their cottage, which enables the cottager to keep a cow or two, and pigs. The poor here may be said to be industrious: they may be often seen themselves, or their children, collecting the horse dung, casually dropped on the roads, for their gardens, or to sell.

SECT. VI.—POPULATION.

The population of this county has increased very much of late, as is evident as well in the towns as villages. In the Appendix, No. XII. will be seen the numbers of births, burials, and inhabitants, collected by Sir R. Sutton, through the hands of several friends, with as much accuracy as could be done.

CHAPTER XVI.

Obstacles to Improvement.

GOOD drainage being one of the greatest improvements in agriculture, the greatest obstacle to it appears to be the bad regulation, or at least defective execution, of the Laws of Sewers; by which a proper outfall is not secured, and is a subject which well deserves revising.

I have before (under the article of water meadows, a very capital improvement of which this county affords great opportunities) suggested the obstacles which occur to it, and a remedy.

Some persons have considered tythes as a great obstacle to improvement; and a law to compel a general compensation for them, as a money, or corn rent, as a remedy.

I must, however, beg to offer my doubts, as to the propriety or efficacy of it.

The right of tythes in the clergy, or lay impropriators, is as much fixed, and guarded by law, as any other property; and, consequently, no alteration should be attempted against their inclination, but for very cogent reasons indeed. It must be allowed, that the taking tythe in kind tends to impoverish the lands of those that pay it, by depriving them of so much straw for manure, whilst it enriches those of the rector, or impropriator, or their lessee. It may likewise sometimes discourage the growing of some particular valuable crops; though, in that case, the rector will generally find it in his interest to come to a composition. The legislature has, indeed, interfered; and, for the encouragement of valuable crops, fixed a certain sum, in lieu of tythes, as in the case of madder.

But what weighs most with me is, that in this, and, I believe, most other counties, more tythes are paid by composition, than in kind. These compositions, from the desire of Clergymen to live well with their parishioners, and partly perhaps from habit, are much lower than the real value of the tythe. If, therefore, a general compensation is to be fixed by law, which must necessarily be by understanding persons upon oath, I apprehend much the greater part of the occupiers would, instead of being relieved, find themselves charged with a much heavier expence than before; and, consequently, instead of a general satisfaction, a general complaint would ensue.

CHAPTER XVII.

Miscellaneous Observations.

THERE is no Agricultural Society appropriated to this county. Many gentlemen, particularly in the northern parts of the county, are subscribers to the Agricultural Society at Doncaster, in Yorkshire.

The weights and measures used in this county purport to be the Winchester bushel and standard weights, but measures in farmers hands often exceed this, and where that is the case, they get a price accordingly; and from dealers being acquainted with their own markets and customers, little subject of complaint, I apprehend, arises on this account.

CONCLUSION.

HAVING already, under the proper heads, suggested the capital improvements of drainage, and water meadows, as also of planting land unfit for other purposes, and of methods which I have ventured to recommend on the latter head, I have now only to add, that I am happy in being able to say, that of late years a great spirit of improvement has arisen in this county, not only amongst gentlemen and considerable farmers, but also amongst the inferior ranks, who begin to have their eyes opened by example, and in many instances have been ready to leave the old beaten track, and adopt better practices in agriculture.

I am sensible, that in the execution of this business, which my zeal for the promotion of agriculture, and the solicitation of a friend, and not any confidence in my own abilities or judgment, induced me to undertake, I may have been guilty of many omissions, and perhaps inaccuracies, which I hope will be thought excusable, when the information was to be collected from many different quarters, and which I hope to see supplied and corrected by the knowledge of others.

I cannot leave the subject without expressing my sense of the liberality and openness of communication, which (except in the case of one or two individuals) I have experienced from the noble and other persons to whom I have applied for information, and to whom I beg leave to return my most sincere thanks.

APPENDIX.

APPENDIX. No. I.

Copy of a Letter to Sir RICHARD SUTTON, Bart.

SIR,

ABOUT a fortnight ago, Mr. Sherbroke desired me to send to you, or Mr. Robert Lowe, my account of the rain, at West Bridgford. On the first of August, 1793, I put down my gauge, which has been accurately attended to since.

In the years 1794, and 1795, I procured by different correspondents, an account of what fell in different places, as you will perceive by the two inclosed cards. I have not yet printed a card of what fell in 1796; being waiting in hopes of receiving an account of what fell in Devonshire or Cornwall, and London. Dr. Campbell's measurement of the rain at Lancaster, for 1796; is, I understand in inches and twelfths, and I have not yet had opportunity of reducing it into tenths.

If I can give you any further information, or explanation of what you will receive with this, I shall be very glad to do it, if you will take the trouble of letting me know it.

I am, with much respect,

SIR,

your obedient and faithful servant,

West Bridgford,   W. THOMPSON.
August 17, 1797.

Quantity of RAIN which fell at West Bridgford in the last five months of the year 1793.

August	4,67
September	2,95
October	0,90
November	2,66
December	1,92
Total inches	13,10

Quantity of RAIN which fell at the following places in the year 1794.

	London.	West Bridgford.[*]	Lancaster.	Kendal.
January	1,64	0,58	3,00	7,29
February	0,94	1,68	6,16	13,47
March	1,26	1,33	3,37	4,51
April	1,52	2,80	3,66	4,18
May	2,54	1,15	2,08	1,99
June	0,50	0,10	2,16	1,45
July	0,62	3,10	2,66	4,16
August	2,42	2,66	3,75	5,34
September	3,71	3,00	6,83	7,67
October	3,36	3,65	7,94	7,32
November	4,44	4,02	3,83	6,01
December	0,37	2,20	5,33	6,20
Total inch.	23,32	26,27	50,81	69,65

[*] In Nottinghamshire.—At Langar, in the same county, the Fall was 29,62.

OF NOTTINGHAMSHIRE.

Quantity of RAIN which fell at the following places in the year 1795.

	London.	West Bridgford.	Lancaster.	Kendal.
January	0,47	1,60	4,00	0,98
February	2,55	1,80	2,21	5,41
March	1,74	1,60	3,02	4,30
April	0,49	2,10	3,33	3,75
May	0,27	0,90	1,50	1,51
June	3,33	3,07	3,75	4,73
July	1,40	1,75	2,47	2,69
August	1,85	1,92	4,58	6,10
September	0,08	0,54	1,06	1,05
October	2,53	4,95	7,08	7,14
November	2,42	3,02	8,50	10,27
December	0,97	1,39	7,48	10,00
Total inch.	18,15	24,64	48,98	57,98

Quantity of RAIN which fell at the following places in the year 1796.

	West Bridgford.	Langar.	Kingston upon Hull.	Lancaster.	Kendal.
January	1,91	1,845	0,91	4,11	2,3688
February	1,34	1,149	1,21	2,6½	3,7764
March	0,50	0,571	0,92	1,9	2,0578
April	1,24	1,169	0,95	2,3¼	1,7646
May	3,32	3,645	3,49	3,4	4,8388
June	0,61	0,709	1,84	0,10¾	2,6112
July	2,87	2,844	4,42	5,7¼	6,6110
August	1,27	1,221	1,15	1,3	1,8936
September	1,70	1,660	1,24	4,3	4,6752
October	1,06	1,206	1,08	3,9	5,9580
November	1,34	1,610	3,70	2,10	1,7142
December	2,00	1,583	2,07	3,0¼	1,9100
Total inch.	18,16	19,212	22,98	37,4½	45,2496

Quantity of RAIN which fell at West Bridgford in the first seven months of the year 1797.

January	1,25
February	0,15
March	1,09
April	2,19
May	3,53
June	4,10
July	1,89
Total inches	14,40

APPENDIX. No. II.

SKEGS appear to be the *avena stipiformis* of Linnæus, described by the Botanical Society at Litchfield, in their translation of the System of Vegetables.

Pannicled, calyxes two flowered, awns twice as long as the seed, culm branchy, stipe form.

They are sown on the worst land: sometimes on a lea, sometimes after turnips, often taken as a last crop. On bad land they may produce about four quarters per acre, which are generally about two thirds of the price of oats. They answer to sow on good land, producing fourteen and fifteen quarters per acre. The kernel is reckoned remarkable sweet good food for horses. They are sometimes threshed, sometimes cut in the straw. They are chiefly grown about Carberton, and will grow where nothing else will.

APPENDIX. No. III.

A Statement of the Cultivation, and other Improvements in CLUMBER PARK, NOTTINGHAMSHIRE, *in the Year 1793, belonging to His Grace the Duke of* NEWCASTLE, *containing about* 4,000 *Acres.*

COMMUNICATED BY MR. MARSON, MANAGER OF HIS GRACE'S IMPROVEMENTS.

IN TILLAGE

	A.	R.	P.	A.	R.	P.
Wheat	67	1	29			
Rye	65	0	29			
Barley	76	1	23			
Oats	210	0	3			
Pease	5	2	0			
Turnips	195	2	0			
Buckwheat						
Total Tillage,				620	0	4

IN PASTURAGE.

	A.	R.	P.	A.	R.	P.
Clover	75	3	1			
Grass for mowing and pasturing	1255	3	31			
Water Meadow	100	0	0			
Total in Pasturage,				1431	2	32
Wood in sundry plantations				1848	1	4
Water in the lake and river below Hardwick				100	0	0
				4000		

APPENDIX. No. IV.

PRIVATE INCLOSURES FROM THE
FOREST AND BORDERS.

In Babworth.

		ACRES
Morton Grange,	Hon. R. Lumley Savile's,	1080
Little Morton,	- - - - - - - - -	600
	G. Mason, Esq. - - - - -	400
Remainder of Babworth, - - - - - - -		2000
Blyth Law Hill,	C. H. Mellish, Esq. - - -	1000
Bilby, - - -	late Morgan Vanes, Esq. - -	260
Scrofton, - - -	Robert Sutton, Esq. - - -	750
Orburton, - -	F. F. Foljambe, Esq. - - -	600
Carberton, &c. -	Duke of Portland (600 planted) about	2400
Bilsthorp Commons,	Hon. R. Lumley Savile's, - -	180
Bothamsell, - -	Duke of Newcastle, - - -	1016
Walesby Warren, in		
Haughton Warren	Ditto - - - - - - -	280

APPENDIX, No. V.

Extract of a Letter to Sir R. SUTTON, *Bart.*

I AM so bad a draftsman, that I am quite ashamed to send you the inclosed draught of *the Cultivator*; but hope you will, by the annexed description of the dimensions, be able to understand it. You see the teeth intersect, and as they are but twelve inches from each other, and by intersecting, the distance is reduced to six inches, and then the breadth of the shares being full three inches, reduces the intermediate space to so small a dimension, that the whole of the ground is entirely broken up, and answers the purpose not only of ploughing, but harrowing likewise, without cutting the quick grass roots in two, which is an advantage that ploughing has not.* It likewise, from the standing forward, and bend of the teeth, brings all the roots up to the top of the land, which is another advantage that cannot be had from the plough. The reduction of labour is another advantage that belongs to this instrument, as four horses and one man will do from six to seven acres per day in sand land. If any other information is wanting that is in my power to give, I shall always be happy to hear from you, and am,

  with the greatest respect,
    SIR,
  Your most obedient humble servant,
       W. BOWER.

---

* It gives me much pleasure to find that the idea I transmitted to the Board, in my notes on the Gloucestershire Report, is reduced to certainty by the instrument described by Mr. Bower. By arranging the teeth in the common harrow, the teeth in each row answering to the interstices of the bars on either side of it, if the clods pass through the interstices of one bar, yet they are certain of being broken by the teeth, in one of the other bars.

          Mr. W. FOX.

References to the annexed plate of the CULTIVATOR.

From A to B the length of the first Bull, 4 feet 6 inches.
From C to D the length of the second Bull, 3 feet 9 inches.
From A to C 16 inches, teeth 2 feet long, and bent near the bottom for the share part to lie flat on the earth, and 1 foot from each other.
From E to F the length of the beam, 6 feet long.
From G to H the length of the iron axle-tree for the small wheels, 1 foot 6 inches.
From I to K the length of the iron that shifts through the beam and fastens with a screw at L, 2 feet.

APPENDIX. No. VI.

List of Open and Inclosed Townships in the
CLAY DISTRICT NORTH OF TRENT.

OPEN.

Wheatley, N.,
Walkringham,
Bole,
Treswell,
Rampton,
Sturton in part,
Headon,
Upton,
E. Drayton,
Dunham,
Ragnall,
Askham,

East Markham,
Ompton,
Normanton and Gresthorp
Sutton
Egmanton
Laxton
Norwell
Kirton.
Oxton
Kneesall
Kirklington in part
Southwell in part

INCLOSED.

Leverton, N. and S.
Beckingham,
Claworth
W. Burton
Laneham
Stokeham
Darlton
Saundby
S. Wheatley
Fledborough
Ossington
Maplebeck
Winkburn
Hockerton
Southwell in part

Weston
Tuxford
Halloughton
Thurgarton
Gonalston
Loudham
Lambley
Gedling
Caunton
Bulcote
Arnold
Calverton
Upton near Southwell
Woodborough

N.B. By the New Inclosures since the first Report was published, the balance between the open field and inclosed townships is altered.

APPENDIX. No. VII.

Open and Inclosed Townships in the VALE *of* BELVOIR.

OPEN.

Cropwell Bishop   Elton

INCLOSED.

Stragglethorp   Kinoulton
Langar   Orsthorp   ⎫
Barnston Basset Flawborough   ⎬ Old inclosure.
Colston   Kilvington   ⎭
Tithby
Elston

INLOSED WITHIN TWENTY YEARS.

Syerston,   Thoroton,   Aslackton,
Flintham,   Carcoulston,   Orston,
Sibthorp,   Newton Old Wark,   Cropwell Butler,
Shelton,   Bingham,   Hickling,
Cotham,   Scarington,   Stanton,
Hawksworth, Whatton,   Granby and Sutton.
Shelston,   Kneeton,

NOTTINGHAMSHIRE WOULDS.

OPEN.

Plumptree,   Wysall
Widmerpool,   Clipston

INCLOSED.

Stanton   Normanton, with some woulds.
Willoughby, in hand.
Over Broughton.

APPENDIX. No. VIII.

THE following account of the extent, jurisdiction, and officers of the Forest were communicated to me by Hayman Rooke, Esq. well known to the literary world.

'The Forest of Sherwood is the only one that remains under the superintendance of the Chief Justice in Eyre North of Trent, or which now belongs to the Crown in that part of England.

In a survey of 1609, it is described as divided into three walks, called North Part, South and Middle Part.

North Part contains the towns of Carberton, Gleadthorpe, Warsop, with Nettleworth, Mansfield, Woodhouse, Clipstone, Rufford, and Edwinstow; the Hays of Birkland and Bilhagh, towns of Budby, Thoresby, Peverelthorp or Palethorp, and Ollerton.

Middle Part, town of Mansfield, Plesly Hill, Skegby, Sutton, Hucknall, Fulwood, part of Kirkby, Blidworth, Papplewick, Newsted, part of Linby and part of Annesley.

South Part, town of Nottingham, part of Wilford, with Radford, Sneinton, Colwick, Gedling, Stoke, Carleton, Burton, and Bulcote; Gunthorp, Caythorp, and Lowdham; Lambley, Arnold, Basford, Bulwell, Beskwood Park, Woodborough, Calverton, and Sauntesford Manor.

FOREST OFFICERS.

Lord Warden, Duke of Newcastle, by letters patent from the Crown during pleasure.

Bowbearer and Ranger, Lord Byron, by the Lord Warden during pleasure.

Four Verdurers elected by the Freeholders for life,

Sir Francis Molyneux, Bart.
J. Litchfield, Esq.
Edward Thoroton Gould, Esq.
William Sherbrooke, Esq.

The verdurers have each a tree out of the King's Hays of Birkland and Bilhagh, and two guineas to each Verdurer attending the inclosure of a break.

Steward, J. Gladwin, Esq.

Nine keepers, appointed by the verdurers during pleasure, having so many different walks.

The keepers have a salary of twenty shillings, paid by the Duke of Newcastle, out of a fee farm rent from Nottingham Castle.

Two sworn woodwards for Sutton and Carlton. Thorney-Wood Chace is a branch of the forest. The Earl of Chesterfield is hereditary keeper by grant to J. Stanhope, Esq. 42 Eliz. The wood and timber of the Crown are under the care of the surveyor-general of the woods. His deputy in the forest is Geo. Clarke, who has a fee tree yearly, and a salary of twenty pounds per annum out of wood sales.

Forest towns, villages, hamlets, or lands belonging to them included in the Sand and Gravel District.

Carberton, Gleadthorpe, Warsop, with Nettleworth, Mansfield, and Woodhouse; Clipston, Rufford, and Edminstow; Budby, Thoresby, Peverelthorp, or Palethorp, Ollerton, part of Kirkby, part of Papplewick, Newsted, part of Nottingham, part of Radford, part of Basford, part of Bulwell, part of Arnold, and part of Calverton.

APPENDIX. No. IX.

Extract of a Letter to Sir RICHARD SUTTON, Bart.

SIR RICHARD,

I AM extremely sorry I have not had an earlier opportunity of communicating these few scattered hints which I have been able to collect, for Mr. Lowe in his survey of this county; and I regret that I have not the pleasure of giving more information upon so important and interesting a subject as I could wish, having been more in the habit of observation than practice; however such as have occured to me worthy of any degree of notice, I here offer you.

Lands under tillage in this county, whatever may be their rotation of crops, are generally fallowed the third or fourth year from the preceding fallow; and it is, and has been the custom, to lay upon such fallows the manure arising from the lands in tillage upon the same farm, and to plough in the same. This practice has been exploded by many, yet discontinued by few. Where it has been the case, that instead of ploughing in the manure with the fallows, and suffering its most subtile parts to descend below the sphere of vegetation, it has been spread upon the stubbles of the succeeding crop, whether wheat or barley before, or early in the winter, a good pease crop has scarcely been known to fail, which if grown upon barley stubbles will prepare the soil for wheat; notwithstanding it may be asserted, that the crop immediately after the fallow will lose the advantage of its manure. I am justified in this opinion by several observations, and my own experience, that little, or no advantage, and sometimes a disadvantage, to the first crop, will be the consequence of manuring the fallows, particularly upon cold or wet land, as the lumps of litter or dung cannot become sufficiently mixed with the soil, but, on the contrary, will hold water like a sponge. As I have both seen and been materially affected by that practice, it has determined me wholly to abandon that mode, as the land by the other method will receive the same nourishment, though at different stages of its rotation.

I have generally observed those crops answer best from their regularity, which are sown upon lands from three to four yards in breadth, and pretty low, provided care be taken to grip and drain them well, to prevent the water standing in the furrows. The only reason I can assign is this: the quantity of rain falling between the ridge and the furrow cannot form itself into such large currents in descent to the lowest parts, as it would on large lands, where the distance from the ridge to the furrow is so great as to suffer the water to accumulate into large streams before it delivers itself, washing up even the roots of the corn, and the best part of the soil. It is evident from practice, that an inclosure, consisting of low small lands, is much easier thrown down, cross cut, and cleansed from weeds, during a fallow, than those larger and higher.

As to that part of the county in which I live, I am sorry to say, that except with a few individuals, I find no general wish for improvement. The hedges neglected, the ditches suffered to be trodden in and grown up, the weeding of crops not attended to, the fallows too little pains taken with; in short, some of the farmers see no farther than the present profits, which often deprives them of future ones. Indeed I perceive an error to pervade the minds of many, as well in other places as here, which would induce them, if permitted, to plough up as much land as possible, and sow four or five crops successively between each fallow; with an eye to immediate advantage, without any thought of leaving their possessions, till in the end their farms become reduced in value; consequently an injury to their landlord, (in which, provided they continue, they must themselves participate) must be incurred.*

For the above reasons, it becomes necessary for the sake of such tenants only, to place all who rent farms under restriction by leases for a year, the heads of which should, before committed to the attorney, be drawn up by some person conversant in speculative husbandry; which leases should, among others, comprehend the following clauses, or articles, viz.

The rent per annum, and times of payment; that the buildings, fences, ditches, drains, &c. be kept in repair; that the timber be not injured; that no swarth be broken without leave in writing by special agreement; that no more than   crops be had between each fallow; that all lands laid down be upon a clean fallow: that the quantity and kinds of seeds be specified: that no hay, straw, or manure, grown or made upon the farm, be sold or taken away; that the landlord shall bear a proportionate share of the expence for all *foreign manure* purchased for the use of the farm; that none of the articles may be broken or varied, without leave in writing, and that not to affect the original agreement, except in that particular case.

Hop ground, of which there is much in the neighbourhood wherein I live, I am under apprehension (generally speaking) tends to injure the lands, in the vicinity thereof, as the whole attention of some growers is to accumulate all the manure possible to be collected for the hop-yard,

---

* I am sorry to say, some land-owners are so jealous of any profit accruing to the tenant, that they are constantly enquiring into his profits and without considering his losses, expences, &c. &c. by advancing his rent on the least suspicions of advantage, he is driven to the waste and destruction of his farm for his own present support.

and totally neglect every thing else; this I judge to be a principal bar against improvement, however the favourers of hop-planting may deny it. For the hop is known to require much and constant support, without repaying any thing to the dunghill; and without considerable assistance there can be no expectation of a crop—lucrative if obtained—and ruinous if otherwise (to some small farmers or cottagers). The tenant (if such) exerts his utmost abilities to contribute all he can to that vegetable lottery, which, if the expression may be permitted, may turn out a sorry blank, or a prize of much greater magnitude than all the profits of the other part of his farm, even in its greatest state of improvement. I am ready to allow that some hop-planters have their other lands duly attended to; that they purchase an additional portion of manure for their plantation, leaving for the farm what it produced. This granted does not destroy the proposition; for either his ground, who sells the manure, or that of some one else, who probably might purchase it for corn or grass land, is thereby deprived of its use; and as the hop-planter can afford to give more per load for it than any other person, it becomes of more than twice the value in such a situation, contiguous to hop plantations, it would elsewhere. This intelligence I do not impart with any view to detract from the value of the science, if such it may be called, or to hinder the growth of the hop, being a very useful article in life; but as I am convinced the observation is founded on fact, I cannot help submitting it to your consideration, having much lamented any deprivation of manure which grass or corn land may suffer; the bare mention thereof leads me involuntarily into a descant upon the use of straw, in thatching houses, barns, &c. as another bar to improvement. For, was the thatch thus expended, applied to the purpose of littering cattle, instead of remaining perhaps twenty or thirty years a dangerous, ill-looking, vermin-harbouring, dusty, and unprofitable covering for a house, stable, or barn, when a handsome covering of pantile might be laid thereon, free from the above inconveniences, and nearly at the same expence; I repeat, was such straw converted into manure, and applied to the grass or corn land of such farms as are thus deprived, the effects would be profitable and lasting: for I consider the additional manure laid upon land, not to have finished its task when it has produced one, two, or three crops, but presume that the increase of such crops contain as much materials for another dunghill as that which produced it, and so on ad infinitum.

If every farmer would seriously consider the above observations, he would hoard every thing that might be converted into this grand primum mobile of agriculture, as gold. I do assure you, I myself am so much a slave, to the thirst after it, that I am sorry to see my servants make use of too large a whisp for the kitchen fire.

Upon my little inclosed farm, little of which is fit for turnips, till of late in grass, I began with manuring such land upon the swarth as I intended to break up in the course of three or four years, at the rate of fifteen or sixteen loads to the acre, during which time I find the pasturage and hay crops much improved, and I doubt not an advantage will be again found when in tillage. I do not, for the sake of a first crop, intend to pare and burn the swarth, and thereby destroy the best part of the soil; but to pare as thin as possibe, and lead the sods on heaps, by means of a *broad-wheeled* cart, to rot down, with a mixture of long litter and other things, as a valuable provision for future wants. The land thus cleared will be more certain of its first and second crop, than if the swarth were ploughed in, as is often the custom, to the injury of all the crops grown before the sward is rotten, especially when it contains the roots and blades of old sharp woodland grass, which is found to occupy cold clay land when long laid down; for the sods are apt to lie hollow and contain much water, which starve the roots of the grain, and has to my knowledge deprived many of their crops. It seems to me a convincing proof, for I have observed many old inclosures so ploughed yield very poor crops for two or three years, after which they have produced very fine ones.

It is my intention, in March or April next, to sow upon an acre of land in the centre of a large grass field, about fourteen pounds of white clover seed. The close was well manured from the fold-yard in November last, and has never yet been harrowed. After sowing and harrowing with a common harrow, to scratch the swarth a little, I mean to make it fine by means of a thorn harrow, and wait the result of my experiment, which is intended to shew how far grass land may be improved without ploughing. Should it succeed, I shall take pleasure in communicating my success to you.

In laying down land for grass, I have often noted, that the white and red clover has been sown together, under a mistaken notion that the red would produce a burthen to compensate for the shortness of the white. The consequence has been this, the red being much stronger and larger, over-runs, smothers, and eats out the white for two or three years, or till its pipy stalks become large, are cut or wounded, so as to take in a sufficient quantity of water to destroy the root; when it dies, a large chasm of bare land appears, and so many upon the field, as bears proportion with the quantity of seed sown with the white. I have remarked that those vacuums are not speedily filled with other grass, unless the land be subject to a spontaneous production.

Gypsum or white plaster threshed in an unburnt state, I have tried as a manure upon grass land, about three years ago, according to the rules prescribed by a pamphlet published concerning the use and good effects of that mineral in America; and am convinced, whatever virtues it may possess in America, it has none here; at least the native plaster which I made trial of had not. I began by laying on six pecks to the acre, and increased the quantity till I expended six bushels per acre, yet no visible alteration of the herbage could be discovered even to this day.

Small inclosures of grass land, in my opinion, answer much better than the same quantity in large ones, (except in pastures for sheep, which are required to be large for the benefit of cool breezes in summer, and less troublesome on account of the fly.) I have long contended this argument with the advocates for large fields, and am glad Mr. Robertson, in his Survey of Mid Lothian, agrees with me in that opinion. I constantly see small closes, surrounded with good white-thorn hedges, bear a greater burthen of herbage in proportion, than large ones; and the reason seems evident. Independent of the manure deposited in the shade by cattle when sheltering, in spring, when the cold sharp winds blow off from the surface of the ground, the warm atmospheric air occasioned by the reflected rays of the sun, in the same manner they blow off the circumambient warmth, caused by perspiration, from our bodies, and render us more sensible of cold than in still calm weather, though the thermometer points the same degree. Thus wind, by having to pass the interwoven branches of the thorn hedge of a small inclosure, is not able to resume its former violent current before it becomes again broken and divided by another of the same fences. In summer, when much hot and dry weather prevails, the hedges shade off the sun and wind, so as to prevent the moisture left by showers and dews from exhalation; of consequence vegetation is more encouraged than where the ground is more parched.

The breed of stock has not yet been much attended to in this neighbourhood; and though I have long wished to improve my own breed of beasts, other business has prevented me from taking that pains necessary for accomplishing so desirable a purpose. The short horned Yorkshire Holderness cow is the most esteemed here, both for milking and fattening. Perhaps Mr. Turnell, of Stokeham, has applied himself more to the breeding of beasts, than any person in this part of the county. As to sheep, having myself been so little conversant with them, I must beg leave to refer you to others better informed.

I am sorry I cannot give you so much intelligence upon the dairy as I wish to have done, not having had sufficient early notice of the enquiry, as it would require a due observation for some years, to come at the production of a dairy farm, people not being in the habit of either keeping clear accounts or burthening their memories with the profits or losses incident thereto. At Fledbro the farms are not very large, perhaps from eighty to one hundred and fifty acres in a farm; the principal occupation is confined to the dairy, though some of them feed very good bullocks thereon. The most correct account I can make out, respecting the production of cheese is, that one cow will produce about three hundred pounds weight of cheese, upon an average, during the summer season.

Calves are much better fed upon linseed pottage mixed with new milk, in the proportion of one third of good mucilage to two-thirds milk; they thrive much better, rest a great deal, and the veal is generally finer; at least the butchers who purchase mine thus fed, tell me so. I have also bred calves in the same manner, only with this difference, when three weeks old, we give them old milk instead of new, and the same quantity of the linseed, viz. one third. The linseed is put into cold water, and heated over a slow fire (one pint to two gallons) for two or three hours, scarcely suffered to boil, then passed through a hair sieve. Warming the linseed saves the trouble of warming the milk.

Having been much employed as surveyor and commissioner for inclosures, you may expect much information on that subject. The improvements are as various as the circumstances under which we find the lordships to be inclosed; the difference proceeds from the disproportion of their soil, quantity of commons, goodness of, or impracticability of making good roads without an enormous expence, contiguity to markets, &c.

A lordship chiefly consisting of good land, with extensive commons belonging thereto, may be said to be the most capable of improvement; a clay one, with a great quantity of common, the next; and a clay one, with scarcely any open common, the least of all; though even the worst, upon an average, will increase about one-fourth in value, after deduction of all expences attending the inclosure, whilst some lordships, under the first description, in a few years have more than doubled their value before inclosure.

We have lands in this neighbourhood, which, I am certain, might be considerably improved by watering, provided the practice was introduced so that the labourers and servants could be instructed in cutting the carriers; but unless a person was constantly attendant he would not have the work done properly. I hope, by means of the Board of Agriculture, and their communications, that practice, as well as other useful ones, will be made general.

You will readily observe, from the irregular and detached way in which I have placed my observations, that I have not had time to arrange them properly; but have put them down as they occured to my mind; this I know your goodness will readily excuse, provided only one hint may prove serviceable to your endeavours; and be assured, should the least benefit be derived from any communication I have the honour to transmit, I shall receive ample gratification in being the means of throwing a mite into the treasury.

Should this or any other of my feeble efforts entitle me to the honour of a future correspondence, I will take the liberty of giving such occasional hints as may hereafter present themselves, leaving the approbation or rejection wholly and alike to the superior judgment of yourself and the Board to determine,

And am,

SIR,

   Your very obedient
    humble servant,
      WILLIAM CALVERT.

Darlton, Feb. 5, 1794.

P. S. I am greatly obliged to your kindness by the perusal of Mr. Robertson's survey, it being a most excellent one in my opinion; as it has afforded me much amusement; the remarks seem made with precision, judgment and impartiality.

APPENDIX. No. X.

OBSERVATIONS *by* Mr. GREEN *of* BANKWOOD FARM, *in* THURGARTON, *on reading the Survey of* MID LOTHIAN, *and applying them to this County.*

Page 48. As it is mentioned that in ground newly brought into tillage, lime has the greatest power; my opinion is the same: but if it do not become an assisting manure upon old tillage, by seeding that old tillage for a few years, as pasturage, when taken up again, I believe the lime then to become of use. Arable lands, in general, are kept too long on the plough, and too little use made of seeds. If practised the contrary way, I believe it would be found an improvement.

Page ditto. Compost lime, road soil, &c. being mentioned as useful, though not a powerful manure; I believe the same: but if more practised in this our country, it would be found very useful to meadow land. It is very proper in that compost to mix a part of dung.

Page ditto. Farm yard dung, I think, to be properly ordered, the shortest time it should lie after being turned in the yard, should be six weeks.——N. B. In the winter, the dung that comes from horses should be regularly mixt with the dung that is made by other cattle, which promotes a greater fermentation than if it was not mixt. Taking dung from the stables by servants in barrows, and turning it down in heaps upon the dunghill, without spreading it with a fork, is an error too much practised by the servant, and too little noticed by the master.

Page 55. In tillage great improvements may be made, by reducing the breadth of the lands, which being from thirty-six to forty-eight feet wide, and elevated in the centre three feet, in such a form, no arable land can be worked to its best advantage; the ridges, at some periods, being over dry, and at others the furrows over wet. It appears also an improvement, at the time of ploughing, to make the furrow become the crown, and crown the furrow, alternately. I have myself tried the experiment upon woodland clay soil, and found it to answer: I wish it was more practised, I think it would be an improvement upon most sorts of soil.——N. B. Upon dry sand soil there is no occasion. It may be said this mode may not suit a damp soil; but as the breadth is mentioned to be such as to suit to be sowed at one cast upon a land, a single hand cast upon a land cannot be sowed regular; but the same land should be sowed by double hand, but with the same quantity as one hand would have done, which places the corn more regular, and likewise prevents it from falling into the hollow furrows, which is too much the custom in sowing, and appears to me an error; for as there is less soil in the furrow, consequently there should be less seed there: and upon moist soil it is but reducing the breadth to such a size of lands as may suit the moistness of the soil. I think that the best means to keep it dry is, that the colder the soil the less the lands should be; and after the seed is sowed, I would have every furrow cleaned out by the plough very carefully; and I think by that method the land would be found in better state than to be put into large and high lands to dry. After a wet winter, high lands are supposed to be firmer and ready earlier for seed; but if the low parts of large lands are readier for seed in the spring, consequently the small lands may be nearly as ready; for to have large and high lands upon wet soil, ridge and furrow in equal order, I think an impossibility. I should give my opinion to have small lands to be the best; but the moister the soil the shorter the lands should be, with proper drainage at the ends; and if not sufficient, a few under drains should be made.

Page 61. Mention is made, that wheat is very seldom sowed after barley, because it is attended with ill effects; for however well it may look in the spring, it falls off very much before it comes to the sickle. I have tried it in Nottinghamshire, which answered as ill as here mentioned. It looked well till got into ear, and instead of filling forward to make a good crop, of which, at that time, it had all the appearance, it dried up at the root, and wasted away, so as to be of very little value. I believe it to be a received opinion with many farmers, that it is an error to sow wheat after barley: I myself am convinced of it.

Page 82. As it is observed, there is a want of spring keeping betwixt the end of turnip-keeping and the succession of seeds, which come one month sooner than the other grass; I cannot say how the seeds are treated in that country; but can say the less they are eat, the earlier they will be upon all sorts of soil that they are sowed upon. To have seeds in the forwardest perfection, they should not be eat at all after the harvest crop is taken, except a little time after harvest, if the ground be dry; and as soon as wet comes, or winter approaches, should be entirely cleared of stock for the winter season. If it is convenient to the occupier of that land to lay a part of dung upon the seeds, while laying dormant, it would be of infinite use, and likewise forward the seeds in the spring. It is to be considered, that in the North of Scotland, they are much colder than we are in Nottinghamshire; but in this county I have tried the above-mentioned upon very cold woodland soil, and frequently have had the succession of seeds immediately to follow the winter-keeping. I have reason to think the backwarder the land, the less the seeds should be eat; and upon very late land, I think should not be eat at all till in the spring pasturage.

APPENDIX. No. XI.

COTTON MILLS IN NOTTINGHAM.

Dennison and Co.
Green and Co.
J. James
Cox and Co. stands still
Hippinstall
Pearson and Co. stands still
Morley
Harris

NOTTINGHAMSHIRE.

Stanford and Burnside. Mansfield
      Ditto   Ditto
Stanton's
Robinson's upper mill ⎫
    Old mill   ⎪
    New ditto   ⎬ Papplewick and Linby
    Middle ditto   ⎪
    Forge ditto   ⎪
    Nither ditto   ⎭
Unwin's, Sutton in Ashfield
Also a mule factory at Sutton
Co. of hosiers, Radford
Thomas Caunt and Co. Southwell
Mr. Chambers, Fiskerton
Handley, Sketchley, and Co. Newark
Late Hardcastle and Co. mule factory, Newark, stands still
Burdin's, Langworth
Hall's and White, Basford
Salmon, Chlevell (nearly finished)
Walsh, Bulwell, part built, but remains unfinished
Rod and Co. Worksop
At Gamston, near Retford, a mill which occasionally spins worsted, cotton, and bump
Lambert's, Loudham, and Gonalston
Worsted mills
Bagshaw, Mansfield
Toplis, Cuckney
Davison and Co. Arnold
Retford mills

APPENDIX. No. XII.

BIRTHS and BURIALS in NOTTINGHAMSHIRE,
for five years, from the beginning of 1789 to the end of 1793, and number of inhabitants.

	Births.	Burials.	Inhabitants.
Annesley	56	36	330
Arnold	366	224	
Askham	38	27	156
† Averham	40	14	190
Babworth	52	16	253
Balderton	121	58	
Barnby in Willows	26	12	
Barton	54	16	
Basford	338	129	
* Beckingham	51	31	400
Beeston	161	66	
Bilborough	44	8	
Bilsthorp and Rufford	30	11	333
Bingham	144	89	
Blyth and Hamlets	198	133	1429
Bole	24	18	148
Bothamsell	38	29	218
* Bleasby	34	26	225
Boughton	34	17	150
Bradmore cum Bunny	96	49	
Bramcote	57	31	
Bridgford, E.	82	44	
Bridgford, W.	45	43	
Bulwell	312	98	
* Blidworth	94	31	435
Burton Joice cum Bulcote	99	45	
Burton, W.	13	8	69
Budby	23	13	72

	Births	Burials	Inhabitants
* Calverton	110	68	
Carberton	8	9	80
Carcolston	18	10	
† Carlton on Trent	52	13	212
Carlton in Lindrick	83	45	573
* Caunton	69	39	310
Chilwell cum Attenborough and Toton	134	77	
Clareborough and Hamlets	208	89	1206
Claworth	63	22	410
Clifton by Nottingham	76	33	
Clifton, N. and S. (including Hasly and part of Spaldforth)	85	58	421
Clipston	26	11	90
Coddington	67	43	
Collingham, N.	71	37	
Ditto, S.	53	34	
Colston Bassett	35	16	
Colwick	9	11	
Cortlingstock	36	18	
Cossam	60	30	
Cotgrave	136	71	
Cotham	8	5	
* Cropwell Great	51	31	
† Crumwell	25	12	188
Cuckney, with Norton Langwith Holbeck Woodhouse	213	103	1534
* Darlton	22	7	137
Drayton, E.	36	22	236
Drayton, W.	17	11	96
* Dunham	54	22	260
Eakring	70	41	400
Eastwood	115	56	
* Eaton	28	23	273
Edwalton	13	9	
Edwinslow	74	39	276
* Edingley	49	19	232
Egmonton	32	20	300
Elksley	42	27	272

	Births	Burials	Inhabitants
Elston	34	25	
Elston Chapel	32		
Elton	10	4	
Eperston	69	39	
Everton	107	80	560
*Farnsfield	91	40	526
Farndon	87	53	
Finningley cum Aukley	95	46	619
Flawborough	6	2	
Fledborough	8	6	78
Flintham	53	41	
Gamston	42	37	371
Gedling, Carlton and Stoke Bardolph	247	151	
Gonalston	27	12	
Griesley	431	223	
Gotham	73	37	
Granby	63	28	
Gringley	73	32	516
Grove	16	10	114
*Halloughton	18	9	80
Harworth with Hamlets	63	42	430
Hawton	19	7	430
Heyton and Hamlets	27	23	229
*Halam	54	18	281
Hawksworth	30	10	
Headon cum Upton	53	33	286
Hickling	43	37	
Hockerton	15	15	115
†Holm	17	11	122
Holm Pierrepont	23	9	
Hoveringham	49	32	
Hucknall Torkard	214	128	
†Kelham	30	29	192
Kingston	20	14	
Keyworth			
Kneesall, Ompton and Kersall	73	42	465
Kilvington	8	4	
Kinoulton	45	19	
*Kirklington	21	4	175

OF NOTTINGHAMSHIRE. 176

	Births.	Burials.	Inhabitants.
Kirkby	107	77	891
Kirton	28	15	165
Kneeton	9	10	
Lambley	70	45	
Langford	14	9	
Laneham	45	24	273
Langar cum Barnston	45	23	
Laxton			480
Lynby	81	45	432
Leak, S.	50	35	
Leak, W.	14	16	
Lenton	113	79	
* Leverton, N. Hablesthorpe and Coates	69	41	376
Leverton, S. and Cottam	73	43	392
Littleborough	11	5	69
Lowdham cum Gunthorp	147	68	
Maplebeck			150
Markham East	106	72	619
Markham W. and Milnton	39	13	180
Mansfield	826	636	6000
§ Marnham cum membris Skegby and Grasthorp			287
Mansfield Woodhouse	154	94	900
Mattersey cum Thorpe	49	33	351
Missen	83	53	557
Mesterton and Stockwith	145	111	1051
* Morton	11	11	79
† Muskham, N.	77	56	408
† Muskham, S.	31	28	261
Newark	1101	756	
§ Normanton			263
Normanton on Soar	39	12	
† Norwell	79	53	476
‡ Nottingham	4603	3615	25,000
Nuthall	31	39	
Ollerton	75	45	291
Ordsall	83	71	504
Orston	61	36	
Over Broughton	33	23	

	Births	Burials	Inhabitants
Owthorp	23	15	
* Oxton	95	58	725
Papplewick			
Palethorp	11	7	40
Plumptree	76	42	
Radford	122	97	
* Ragnall	27	21	151
* Rampton	41	35	309
Ratcliffe on Trent	119	61	
Rempston	36	13	
Ratcliffe on Soar	15	5	
Retford, E.	347	248	1910
Retford, W.	91	102	524
† Rolleston	38	44	373
Ruddington	148	71	
Sandby	14	15	72
Scarrington	22	18	
South Scarle cum Besthorp	36	22	
Screveton	38	21	
Selston	117	71	524
Scrooby	36	21	209
Shelford	70	39	
Sibthorp	7	6	
Shelton	16	6	
Skegby	63	47	180
Stanton on the Woulds	10	4	
Staunton	16	12	
Southwell and Normanton	376	229	2295
Stapleford	120	96	
Stoke East	58	50	
Stokeham	5	2	30
Strelley	55	55	
Sturton and Fenton	100	44	485
† Sutton on Trent	73	11	558
Sutton and Lound	49	33	571
Sutton in Ashfield	483	382	3492
Sutton Bonington	115	79	
Syerston	16	12	

	Births	Burials	Inhabitants
Teversall	35	60	316
Thorpe	7	3	
Thorney, including Wigsley and Broadholm	33	18	214
Thoroton	17	11	
Thrumpton	16	10	
Thurgarton	45	33	
Tithby cum Little Cropwell	70	38	
Treswell	23	21	174
Trowell	89	50	
Tuxford	121	91	800
*Upton	74	32	362
Walesby	37	14	260
Walkeringham	57	46	406
Whatton	83	50	
Warsop and Sookholm	172	94	848
Wellow	66	48	350
Weston	30	34	267
Widmerpole	26	24	
Wilford	74	66	
Willoughby in Woulds	74	30	
Winkborn	39	11	140
*Wheatley, N.	72	45	358
Ditto, S.	3	1	29
Wollaton	142	89	
*Woodborough	85	40	
Worksop and Hamlets	499	257	3008
Wysall	35	11	

Total Baptisms - - - - - - 20,274
Total Burials - - - - - - 13,277
Number of Inhabitants - - - - 115,598

N. B. Where the numbers of inhabitants are set down they were taken from house to house. With regard to the others, the total was brought out, by multiplying the deaths by 41, the lowest proportion of persons living found in any district, (viz. that of ten villages on the Trent near Newark) to the deaths, where the actual numbers living were taken. In about eighty villages near Retford, the deaths turned out about one in forty-five, and in a good many, about Southwell the same. In the district about Mansfield (fourteen parishes) one in fifty-five, owing probably to the number of children, &c. from other counties employed in the mills and manufactories.

---

† In the places thus marked, the totals of births and burials were by mistake taken for six years, but have been reduced to the average of five, and the numbers of inhabitants having been taken from house to house makes it quite immaterial

* In the places marked thus, the births and burials were taken for five years, ending 1792, the returns of 1793 not being then made into the office at Southwell, which, it is apprehended, would make little or no difference, and in most of them the numbers were taken from house to house.

§ In the two places marked thus, the inspection of the register, was by the Rev. Thomas Clark, Vicar of Normanton, and Curate of Marnham; expressly refused to any person coming from Sir Richard Sutton, but the numbers were taken from house to house.

‡ Multiplying the average of burials by thirty-two, according to the paper annexed, marked (B) will not bring out the full number of 25000, but a gentleman who had a great share in drawing that paper, is persuaded that number is not too high, especially as there are congregations of dissenters who do not register; and the dissenters in some villages near Nottingham would somewhat raise the total of the county.

The paper marked B will shew the increase of Nottingham, since 1779.

B

An account of the number of INHABITANTS *of the Town of* NOTTINGHAM, *with the number of Houses and Families, distinguishing each street alphabetically, taken from Monday the 20th, to Saturday the 25th of September inclusive,* 1779.

THE method adopted, and rules that were observed in taking the following account:

No militia man or soldier was reckoned, but their families were numbered if they were housekeepers in the town.

If any part of a family were absent upon a visit or journey they were counted, as were all children who boarded at schools, &c. in the town; therefore no persons upon a visit here, nor children belonging to the town if boarded out of it for education, &c. were taken.

Distinctions were made of houses, families and inhabitants in each parish, that a succinct account of each might be given.

The hospitals were not numbered amongst the houses, but the people were taken as inhabitants.

	Houses.	Families.	Souls.
Angel Row	24	23	139
St. Ann's Street	30	35	149
St. Ann's and Copies	4	5	24
Back Lane and Penny Foot Row	32	35	197
Barker Gate	79	94	418
Bearward Lane	73	83	424
Beast Market Hill	14	14	88
Beck Lane	13	13	69
Beck Lane Hospitals	—	—	13
Beck Barn or Pottery	13	19	87
Bedlam Court	19	20	89
Bellar Gate	46	50	231
Boot Lane, from Parliament Street, to the Joiner's-Arms, exclusive of Kayes's Buildings	110	128	624
Bottle Lane	16	19	93
Bowling Alley Hill	6	6	34
Bridge End, see Hollow Stone			
Bridlesmith-gate and Rose Yard	81	86	415
Broad Lane to St. John's	22	26	143
Broad Marsh and Darker's-Court	68	85	375
Byard Lane and Chappel's-Court	13	13	69
Bilby's Hospital	—	—	18
Butt Dyke, see Toller's Hill			
Blowbladder Street, see Mount Hall Gate			
Carter Lane	46	50	230
Castle Gate	87	97	447
Chandler's Lane	15	23	90
Chappel Bar to Nix's Yard	29	31	145
Cappel's Court, see Bryard Lane			
Cheapside	10	10	49
Chesterfield Lane	18	19	83
Coalpit Lane	20	21	112
Cow Lane	27	28	129
Cuckstool Row	12	13	94

OF NOTTINGHAMSHIRE. 181

	Houses.	Families.	Souls.
Cabbage Court	7	9	47
Charlotte Street and three Salmon's Yard	40	41	224
Collin's Hospital	—	—	42
Drury Hill	13	14	74
Darker's Court, see Broad Marsh			
Engine House and Neighbourhood	9	13	59
Finkhill Street and Walnuttree Lane	26	30	136
Fisher Gate	60	67	292
Fletcher Gate	30	30	145
Friar Lane or Mont Hall Lane	6	6	38
Flint's Court, see Garner's Hill			
Gilliflower Hill and Rock Holes	13	15	69
Goose Gate & Hockley to Parivicini's R.	65	68	304
Greyfriar's Gate	54	60	293
Griddlesmith Gate	43	43	204
Greyhound Yard	30	36	162
Glass House Lane to Charlotte Street	45	59	290
Garner's Hill and Flint's Court	7	9	44
High Pavement	39	39	207
High Street	10	10	75
Hockley, see Goose Gate			
Hound's Gate	81	90	407
Hollow Stone, Bridge End and Malin Hill	50	54	258
St. John's and Keyworth's Houses	15	22	85
St. James's Lane	59	64	326
Jew Lane, see Spaniel Row			
Johnson's Court	14	19	84
Kayes's Buildings in Boot Lane	24	28	95
Long Row, from Nix's Yard to Cow Lane, including all the Yards, except Greyhound-Yard	140	154	824
Low Pavement	18	19	99
Leen Side	44	56	265

AGRICULTURAL SURVEY

	Houses	Families	Souls
Malin Hill, see Hollow Stone			
St. Mary's Church Side	24	25	109
St. Mary's Gate	42	42	178
Marsden's Court	10	11	61
Middle Pavement	15	15	67
Mount Hall Gate or Blowbladder Street	19	20	98
Mount Lane or Middle Hill	13	14	64
Milstone Lane to Beck Barn	44	44	205
St. Mary's Workhouse	—	—	108
Middle Marsh	20	21	94
Narrow Marsh and Long Stairs, including all the Yards, &c.	197	226	1035
New Change and Shoe Booths	13	13	64
St. Nicholas's Workhouse	—	—	57
Parivicini's Row, Owen's Court, &c.	27	31	172
Peck Lane	6	6	23
Pennyfoot Row, see Back Lane			
Pepper Street	10	11	42
St. Peter's Workhouse	—	—	42
St. Peter's Church Side	14	14	62
St. Peter's Gate and Church Yard	37	38	169
Pilcher Gate	23	28	120
Parliament Street and Back Lane	275	299	1504
Plumptre's Hospital	—	—	14
Queen Street	8	10	44
Quaker Lane, see Spaniel Row			
Rosemary Lane	12	13	61
Rockholes, &c. beyond Glass House Lane	19	19	71
Rockholes, Gilliflower Hill, see Gilliflower Hill			

OF NOTTINGHAMSHIRE. 183

	Houses	Families	Souls
Spring Gardens, including all the New Houses South of St. Ann Street, and East of Glass House Lane	68	86	365
Shambles, see Smithy Row			
Sheep Lane	17	20	86
Short Hill	17	18	79
Smithy Row and Shambles	18	17	82
Spaniel Row	10	13	59
Stephen's Court, see Leen Side			
Stoney Street	29	30	156
Stoney Street Hospital	—	—	16
Swine Green	17	21	108
Timber Hill	21	23	149
Trent Bridge	9	14	71
Turn Calf Alley	15	18	97
Toller's-Hill and Bal Dyke	93	97	519
Tabernacle Alley, including all the Houses at the Back of Boot Lane, from Parliament Street to Charlotte Street	61	61	284
Walnut-tree Lane, see Finkhill Street			
Warser Gate	45	48	198
Wheeler Gate	38	40	207
Woolpack Lane	54	61	332
White Rents in Houndsgate	—	—	63
Total Number	3191	3556	17584
Brewhouse Yard, an Extra Parochial Place	—	—	127
	3191	3556	17711

A division of the number of Houses, Families and Inhabitants in each Parish:

	Houses.	Families.	Souls.
St. Mary's Parish	2314	2584	12637
St. Peter's	446	497	2445
St. Nicholas's	431	475	2502
Brewhouse Yard	—	—	127
Total	3191	3556	17711

An account of the Burials for seven Years, from 1772 to 1778, inclusive:

St. Mary's	2315
St. Peter's	608
St. Nicholas's	790
Castlegate Meeting	74
Baptist ditto	98
Quakers ditto	18
Total	3903

Houses that are now uninhabited in this town.

St. Mary's Parish	57
St. Nicholas's ditto	10
St. Peter's ditto	9
Total	76

The inhabited houses contain as near as possible, five and a half upon the average to each house.

If we divide the number of inhabitants by the annual average number of burials, we shall discover that it will require about thirty-one years and ten months to bury a number equal to that of the whole town, consequently nearly one in thirty-two of the inhabitants die annually. We shall also find that by ascertaining the number of people and burials, the comparative healthfulness of places may be determined; making proper allowance for those who die in their infancy, and for the extraordinary increase or decrease of the people by acquisition or emigration. This comparison has not been made in many places in England, because the mistaken apprehension of new taxes, and other reasons, make the people jealous of being enumerated.—It his however with pleasure we declare, that we found very few such groundless fears to prevail here; but on the contrary, the generality of the people gave their numbers with great good nature and chearfulness.

It appears that in a very healthful parish, called Holy Cross, adjoining to Shrewsbury, one in thirty-three die annually, though in Shrewsbury and Northampton one in twenty-seven, and in London one in twenty-one; but with respect to London, the computation has been made only from the number of houses. In many places of Europe, regular accounts are annually taken, from which we find at Vienna one in twenty die every year; at Berlin one in twenty-six, but this number would be smaller only for an extra encrease of people of late years; at a country parish in Bradenburgh one in forty-five; and the same in those healthful villages of the Pais de Vaud near Geneva; but this high number may proceed from the emigration of the natives, of which Dr. Tissot, in the Introduction to his *Advice to the People*, very much complains. We must not conclude that because this number is twice that of London, therefore the chance of life to adults is two to one against London; it is only so to the new born infant; hence the necessity in these calculations of always taking into consideration the number of infants that are annually buried.—At Vienna half the number of inhabitants die before they are two years of age; at Berlin two and three quarters, at London three, at Northampton six, at Holy Cross twenty-seven, and in the Pais de Vaud forty-one.

From a comparison of the foregoing premises, it is with peculiar satisfaction that we conclude Nottingham to be a very healthy situation, for we nearly come up to the standard of Holy Cross, and should certainly exceed it, if it was not for the numbers that die here in their infancy; where poor people are forced to neglect their offspring to procure a subsistence, it is no wonder if half of those who are born die young. Dr. Deering in his Antiquities of Nottingham, page seventy-eight, gives us 2331 for the burials in seven years; of which 1072 were infants. Near half therefore die in their infancy, which cannot be the case at Holy Cross, where half that are born live to the age of twenty-seven.—The doctor in the year 1739, enumerated the inhabitants of this place, and making a proportionate allowance for some omissions and deficiencies in his account, it appears there were at that time about 10720 souls in this town; taking also his annual average of burials for seven years, by which if we divide the number of inhabitants, it will appear that nearly thirty-two years was then the requisite time to bury the whole number of the people. This similarity at forty years distance with the present statement, most certainly removes the suspicion of inaccuracy in both accounts. We are aware that the judicious may possibly observe that the very great additional number of people since Dr. Deering's account, will of course operate here as it did at Berlin, and make this place also appear more healthful than it really is, and such would certainly be the case, if during the time that we have given the state of the burials, there had not been a very unusual drain of the inhabitants into the army and militia, which we conceive fully counterbalances the increase.

APPENDIX. No. XIII.

From Mr. GOULD, late Steward to the DUKE OF PORTLAND, on making ponds in dry pastures.

The most approved method of making ponds or artificial pools in Derbyshire.—After removing the earth and forming a proper bason, take a quantity of lime, (ashes of lime will answer the same purpose) and spread the same over the whole surface, about five inches thick; upon this bed of lime lay a coat of well tempered clay, eight inches thick: this must be beat down extremely well with wooden hammers, to prevent the clay from cracking. Upon the top of the clay lay a second coat of lime, of the same thickness as the first: the whole is then paved or pitched with small stones to prevent the cattle from injuring the materials of the pond; the clay will naturally hold up the water, and the lime prevents the worm from striking upwards or penetrating downwards to injure the clay.

APPENDIX. No. XIV.

CURE OF DISEASES IN CORN AND CATTLE.

SMUT IN WHEAT.

The following receipt for preventing the smut in wheat, has been practised these twelve years past.

Take twenty-eight gallons of water, boil in a few gallons of it one pound of arsenick, then mix all together, and steep your wheat in it for six or eight hours; when taken out, mix well with fresh lime as usual. The wheat should be put through a riddle, and what swims at top skimmed off.

FOR THE ROT IN SHEEP.

Take five quarts of boiling water, pour it upon a handful of rue chopped small, and cover down the tea thus made for ten or twelve hours. Then strain it off and add thereto as much salt as will make it swim an egg new laid. Add to it a lump of bole armoniac as big as a pullet's egg, and double that quantity of chalk, both well pounded before they are mixed with the tea; when well incorporated add half a pound of flour of brimstone. The whole well mixed, is a sufficient drink for a score of large pasture sheep. To each sheep, after fasting four or five hours, give half a pint of the mixture in a small horn in three horns full, letting it rest, to take breath and cough, a minute between each; for want of which many have been killed in the operation. Three drinks have been given in various years in the months of September and October, at the distance of a week between each drink, with great success, not only to prevent but to cure the rot in sheep. Whilst one person is administering the medicine, another should be employed in stirring the ingredients well together.

N.B. The sheep should be kept fasting two hours after the medicine. It is adviseable that the person who mixes up the ingredients, as well as he that stirs them together, should use a large wooden spoon, lest by using his hands too freely, the compound should take off the skin.

ANOTHER RECEIPT.

Two ounces diapente in a quart of brandy for twenty sheep, given as a preventative, three times, at Michaelmas, Christmas, and Candlemas.

FOR THE WATER.

To prevent sheep from dropping of the water, take one pound of tobacco, and boil it in seven quarts of water one hour. Then put to it four ounces of saltpetre, one ounce black pepper and two ounces of spirits of turpentine. Two spoonfuls of this mixture are given to the lambs about a month after they are taken from the ewe, and two more spoonfuls about a month after the first. Great care must be taken to shake it well together, before it is given, otherwise the turpentine will swim on the top.

APPENDIX. No. XV.

Letter from Mr. RAYNES *of* Stonehill), *on Cultivation of a Sand Farm.*

SIR,

Agreeable to your request, I inclose you a few hints on the practical part of agriculture, as performed by myself upon a poor sandy soil, much like what you generally find upon the forest; first I begin with a fallow, which I endeavour to clean as well as possible; if an order could be had after harvest it would be of great advantage, but at any rate plough it as soon as you have done your Michaelmas seed time, taking care to turn over all the soil, and leave it as open as you can; as soon in the spring as you think your land will work, begin to harrow or scarify, but I should recommend the latter. The scarifier is an invention of the Rev. James Cooke, Red Lion Square, London, and which is particularly useful in sandy soils; for I have applied it this summer in my turnip fallows, and can say I never had them in better tilth, notwithstanding the wetness of the summer. The scarifier is worked by two horses, which will do eight acres in one day, and after the first ploughing, will answer equal, if not better than a ploughing: this I recommend to be drawn across the fallow, after which make use of the harrows and get off the twitch, then scarify again, drawing it the contrary way from what you did before, when you will soon get rid of your twitch. The scarifier has only one row of teeth; if your fallow is very foul, put in only four teeth; after you have got it tolerably clear, put in the remainder, which will collect all the short twitch; this implement has one very excellent qualification above all the cultivators I have seen, namely, that it does not break or cut the twitch, and this implement by performing so much in one day, gives a longer time between the orders of the fallow, so that the weeds vegetate and are in consequence more effectually destroyed by the succeeding operation; and by this mode of fallowing you will be much earlier than by the plough and harrow only, which will give you an opportunity of manuring early and incorporating the dung well with the soil, by giving the fallow two after-ploughings, and by such incorporation of the dung and soil there will be a certainty of succeeding in the turnip crop. I must not forget to say we are very much indebted to the Norfolk farmers for the invention of a machine for sowing turnips, one of which I have, and for regularity and dispatch it is not to be excelled: the price is about twelve shillings. I find malt combs a very excellent manure for turnips, twelve quarters per acre is a very good dressing, and will more than equal twelve loads of manure produced from the farm yard: after turnips I have barley with which I sow seeds, such as white clover, 15 lbs. trefoil, 5 lbs. and rye grass one bushel, all of the best quality per acre; in the succeeding course I vary my seeds, by sowing red clover with trefoil and rye grass, which I find a complete change from the other; by this method you will be certain to have good seeds in the spring. I avoid as much as possible pasturing the young seeds after harvest; they will be much earlier and more abundant at spring; this I pasture for two years if not wanted for fodder, when I should prefer cutting the second year, on account of the trefoil, which will be totally destroyed by mowing; after which plough it at Michalemas, and sow it with wheat or rye according to the strength of the soil. It is a most excellent method of fallowing from the sward, but yet I think when you can manure for the whole of your fallows, it will be the most profitable way to take a crop of wheat or rye first; another thing, you will fallow much easier after a crop than from the sward, except your land is quite clear of twitch, which is seldom the case in sandy land. I scarce ever mow any seeds for fodder, as it is very injurious to such soils as mine, and believe it very possible to keep horses winter and summer without mowing any seeds which is done by cutting them unthreshed oats or wheat, straw or beans, in the winter months: in summer they are kept in the fold upon green fodder. A certain quantity of land kept in a garden stile for such succession of green fodder is an excellent practice. To let land lay three years in seeds I apprehend is not a good practice, particularly in light sandy land; for the third year's seeds will bear very little stock in proportion to the first and second years; of course the land cannot benefit by the stock, and if it only benefit by rest, it follows that it should be ploughed up at the end of the second year, and to get it into a state of fresh seeds again as soon as possible.

I am,

SIR,

Your obedient humble servant,

Stone Hill, F. RAYNES.
April 3, 1798.

To *Robert Lowe, Esq. Oxton.*

P.S. As clover seed in general is very dear, and from that circumstance land is often not half seeded, I should recommend to every farmer to grow his own seeds of every sort, which may be done in all situations.

N. B. *This Letter not being received till after the corrected Report was printed off, could not be inserted in the body of it.*

FINIS.

www.ingramcontent.com/pod-product-compliance
Lightning Source LLC
Chambersburg PA
CBHW020916230426
43666CB00008B/1473